Multilevel Modeling
in Plain Language

Multilevel Modeling
in Plain Language

Karen Robson & David Pevalin

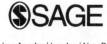

Los Angeles | London | New Delhi
Singapore | Washington DC

Los Angeles | London | New Delhi
Singapore | Washington DC

SAGE Publications Ltd
1 Oliver's Yard
55 City Road
London EC1Y 1SP

SAGE Publications Inc.
2455 Teller Road
Thousand Oaks, California 91320

SAGE Publications India Pvt Ltd
B 1/I 1 Mohan Cooperative Industrial Area
Mathura Road
New Delhi 110 044

SAGE Publications Asia-Pacific Pte Ltd
3 Church Street
#10-04 Samsung Hub
Singapore 049483

Editor: Mila Steele
Assistant editor: James Piper
Production editor: Ian Antcliff
Copyeditor: Richard Leigh
Proofreader: Louise Harnby
Indexer: David Rudeforth
Marketing manager: Sally Ransom
Cover design: Jennifer Crisp
Typeset by: C&M Digitals (P) Ltd, Chennai, India
Printed and bound by CPI Group (UK) Ltd,
Croydon, CR0 4YY

Library of Congress Control Number: 2015938248

British Library Cataloguing in Publication data

A catalogue record for this book is available from
the British Library

ISBN 978-0-85702-915-7
ISBN 978-0-85702-916-4 (pbk)

Contents

Multilevel Modeling in Plain Language is supported by additional resources to aid your study, including Stata and R datasets and command language, which are available for you to download at **https://study.sagepub.com/robsonandpevalin**

About the Authors

Karen Robson is Associate Professor of Sociology at York University. Her research areas include the barriers to postsecondary education for marginalized youth, intersectionality as a policy framework and critical race theory. She also has a strong interest in the analysis of large longitudinal data sets to examine issues around social mobility and the transition to postsecondary education. Dr Robson has also written key textbooks in the area of social research methods and the sociology of education, as well as several articles in various sociology journals.

David Pevalin is Professor in the School of Health and Human Sciences and Dean of Postgraduate Research and Education at the University of Essex. His research focuses on macro and micro social inequalities in health. He co-authored (with Karen Robson) *The Stata Survival Manual* (Open University Press, 2009), co-edited (with David Rose) *The Researcher's Guide to the National Statistics Socio-economic Classification* (SAGE, 2003), and authored research reports for the Department of Work and Pensions and the Health Development Agency. He has published papers in *Journal of Health & Social Behavior*, *British Journal of Sociology*, *The Lancet*, *Public Health* and *Housing Studies*.

ONE

What Is Multilevel Modeling and Why Should I Use It?

CHAPTER CONTENTS

You are probably reading this book because someone – a professor, a supervisor, a colleague, or even an anonymous reviewer – told you that you needed to use multilevel modeling. It sounds pretty impressive. It is perhaps even more impressive that multilevel modeling is known by several other names, including, but not limited to: hierarchical modeling, random coefficients models, mixed models, random effects models, nested models, variance component models, split-plot designs, hierarchical linear modeling, Bayesian hierarchical linear modeling, and random parameter models. It can seem confusing, but it doesn't need to be.

This book is for a special type of user who is far more common than experts tend to recognize, or at least acknowledge. This book is for people who want to learn about this technique but are not all that interested in learning all the statistical equations and strange notations that are typically associated with teaching materials in this area. That is not to say we are flagrantly trying to promote bad research, because we are not. We are trying to demystify these types of approaches for people who are intimidated by technical language and mathematical symbols.

Have you ever been in a lecture or course on a statistical topic and felt you understand everything quite well until the instructor starts putting equations on the slides, and talking through them as though everyone understands them? 'As this clearly shows … theta … gamma ….'. Have you ever been in a course where several slides of equations are used to justify and demonstrate a procedure and you just didn't understand? Perhaps you were thinking, 'These equations must mean something in words. Why can't they just use the words?' You might also just want some practical examples that are fully explained in plain language that you may be able to apply to your own research questions. If this sounds like you, then this book is for you. If you are really fond of equations, then we're afraid that our approach in this book won't appeal to you.

Before we get started, we also want to emphasize that this book is not about 'dumbing down' complicated subject matter; it's about making it accessible. We are not endorsing using modeling techniques without understanding them. This leads to sloppy and unscientific analyses that are painful to read. What this book does is unpack these sophisticated techniques and explain them in non-technical language. We assume that you understand the principles of hypothesis testing, sampling, research design, and statistical analysis up to and including ordinary least squares regression (OLS) with interaction terms. The techniques discussed here are just extensions of regression. Really! We do try to keep the jargon to a minimum but, as with all things new, there are some new terms and phrases to get to grips with.

So, why might you have been told that you need to use multilevel modeling? The chances are that it is because your data have a 'nested', 'clustered', or 'grouped' structure and you are being guided to a (regression) technique that accounts for this type of data structure.

The idea of nesting, or clustering, is central to multilevel modeling. It simply means that smaller levels of analysis are contained within larger grouping units (Figure 1.1). The classic example is that individual students are nested within schools, but nesting can take other forms, such as individuals (Level 1) within cities (Level 2), patients within hospitals, siblings within families, or employees within firms.

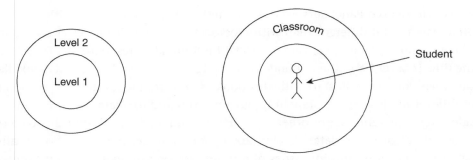

Figure 1.1 Two-level nesting

In these types of two-level models, the lower level consisting of the smaller units (often individuals) is called Level 1, and the next level is referred to as Level 2. You may also have a third (or higher) level within which Level 2 units are nested (see Figure 1.2) such as students (Level 1) within classrooms (Level 2) within schools (Level 3), patients within hospitals within districts, siblings within families within neighbourhoods, or employees within firms within nations.

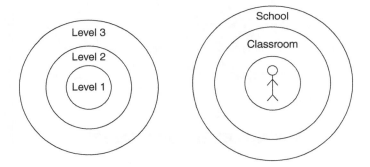

Figure 1.2 Three-level nesting

Multilevel modeling can deal with three or more levels of nesting. However, in this book we will focus on two-level models. As you gain more expertise in multilevel modeling, you may want to explore more complex structures, but for the scope of this introductory book, analysis with two levels will serve as the foundation upon which illustrative examples are created.

This book will also only consider cross-sectional nested data. There is another variation of nesting with longitudinal data. Longitudinal data are obtained when information is collected from respondents at more than one point in time. For example, people are interviewed annually in a longitudinal survey such as the

British Household Panel Survey. The way nesting is conceptualized with longitudinal data is a bit different than with cross-sectional data. Data collection events (Time 1, Time 2, Time 3, etc.) are nested within individual respondents. Therefore, the time is Level 1 and the respondent is Level 2 (Figure 1.3). If your data look like this then you may still start with this book as an introduction to cross-sectional multilevel models before branching out into longitudinal data. On the positive side, there are a number of other potential techniques for analysing this type of longitudinal or panel data. Halaby (2004), for example, offers a sociologically based primer for examining issues of causality in longitudinal data. Fitzmaurice et al. (2012) have written a more detailed textbook on the topic of applied longitudinal data analysis that is targeted at a multidisciplinary audience.

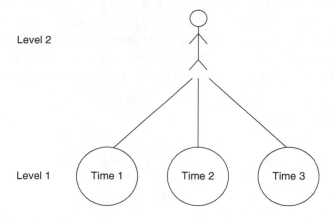

Figure 1.3 Longitudinal nesting

MIXING LEVELS OF ANALYSIS

There are two errors in causal reasoning that have to do with mixing different levels of analyses, which are illustrated in Figure 1.4. The first is known as the *ecological fallacy* and has to do with generalizing group characteristics to individuals. If we analyse the effect that average neighbourhood income has on the crime rates of that neighbourhood, we are comparing group characteristics with group characteristics. To extend this argument to *particular individuals* in the neighbourhood can be misleading. It is not appropriate to apply group-level characteristics to individual-level inferences. We may well find that as average income in a neighbourhood decreases, the crime rate increases – but we cannot say that if an individual's income decreases, he or she is more likely to commit crime!

The ecological fallacy can be demonstrated in a number of ways. Another common misinterpretation of group characteristics is to look at the average income in two very different communities that are about the same size. In Wealthyville,

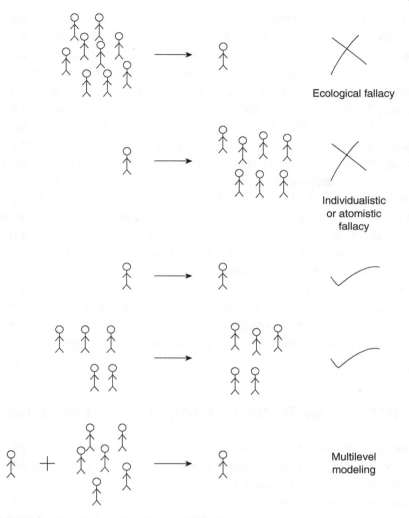

Figure 1.4   Units of analysis and making generalizations

the average household income is $500,000 per year. In Poorville, the average household income is only $15,000 per year. If 10,000 households live in each community, we would say that the average household income across both communities is $257,500 per year. This would give a completely inaccurate representation of the communities, however, because it doesn't represent the household income of *anyone*. It is far too little to represent Wealthyville (just about half the actual household income) and too high to represent Poorville (over 17 times the actual household income). By taking group characteristics and trying to generalize to individual households, we have committed the ecological fallacy.

Keeping your units of analysis comparable also applies to arguments made in the opposite direction – generalizing individual processes to group processes.

This problem is known as the *atomistic fallacy* (or individualist fallacy). People make this mistake when they take results from individual-level data and apply them to groups, where the context may be very different. We may find, for example, that being an immigrant is associated with an increased risk of mental health problems. A policy solution, however, of creating mental health programmes for all immigrants may be misguided, if contextual variables (at the group level) are not taken into account. It may be that immigrants in large cities have better mental health than immigrants in small communities (where they may be isolated) (Courgeau, 2003). If we simply take individual-level characteristics and apply them to groups, failing to take contexts into consideration, we may come to conclusions based upon flawed logic.

Both the ecological and atomistic fallacies are errors that researchers make when they take data at one level and try to make generalizations to another level. As social scientists, we know that individual characteristics (e.g. age, gender, race) and contextual-level variables (e.g. school, neighbourhood, region) are important determinants for many different outcomes of interest. In multilevel modeling, we use both individual and group characteristics and our outcomes can be modeled in ways that illustrate how individual and group characteristics both affect outcomes of interest, and how group characteristics may influence how individual characteristics affect the outcome of interest, given certain contexts.

THEORETICAL REASONS FOR MULTILEVEL MODELING

Your models should always be theory-driven, and the best model choice is one that corresponds to a sound theoretical rationale. One that is often overlooked is the general theoretical arguments around how the social world is portrayed. Education researchers, such as Bronfenbrenner (1977, 2001), have argued that the outcomes of individuals, particularly children, cannot be understood without taking different contexts into perspective. His ecological systems approach identifies a number of different contexts to be taken into consideration in terms of how they work independently and together to create the environments in which children live. By looking at data collected from individuals, we are focusing on the micro-level (i.e. individual) effects of specific characteristics on outcomes of interest, but it is more likely the case that these micro-level effects vary significantly across larger units at the meso (school or community) and macro (municipal or national) levels. The micro and the macro (and the micro and meso) interact. This theoretical perspective is most readily tested with statistical techniques that recognize these important distinctions.

Although discussions in this vein invariably resort to examples from education research, the applications and theoretical motivations apply across a range of disciplines, including health, political science, criminology, sociology, and management research. Scholars from all these disciplines have noted the importance of linking the individual (the micro) and the contexts in which he or she lives (the macro). The popularity of theories that focus exclusively on the individual or solely on higher levels (groups, firms, nations) is being overshadowed by approaches that try to mix the two and presumably give a more accurate depiction of the complexity of the social world.

WHAT ARE THE ADVANTAGES OF USING MULTILEVEL MODELS?

Well, as the name implies, multilevel models are equipped to analyse multiple levels of data. The information about individual and group characteristics is retained and separate estimates are produced for both. Adjustments are made for correlated error terms and for the different degrees of freedom (i.e. Level 1 degrees of freedom and Level 2 degrees of freedom). Perhaps most importantly, we are also able to do cross-level interactions so that we can explain how Level 1 predictors affect our dependent variable according to different contexts (Level 2 predictors). Additionally, we can look at how Level 1 predictors interact with each other and how Level 2 predictors interact with each other. And because multilevel modeling is just an extension of OLS, much of the knowledge you already have about OLS will come in handy for model-building and the interpretation of the results.

We now show that using regular single-level regression techniques to address multilevel issues is fraught with problems. The main problem is that your results are likely to be filled with errors that originate from various violations of the regression assumptions. In many cases, we get poorly estimated results that will be statistically significant, leading us to reject our null hypotheses when we really shouldn't be rejecting them. In other words, finding associations where none exist, erroneous conclusions, and possibly leading to ineffective policies.

STATISTICAL REASONS FOR MULTILEVEL MODELING

There are many statistical reasons to choose multilevel modeling techniques over standard linear OLS regression. You may have accepted that you just need to learn this and don't really care about all the technical reasons why, but we would argue that you should at least grasp the basic reasons why OLS is deficient in estimating models with nested data. You may have already tried some of the 'workarounds', which we discuss below, in OLS to model nested data. These 'workarounds' have been, and continue to be, used by many researchers and it is not difficult to find examples of them in the literature. They are still technically flawed, however, and we explain below why it is problematic to choose OLS, despite these 'workarounds', when trying to deal with nested data structures.

ASSUMPTIONS OF OLS

Multilevel modeling is an extension of OLS. The thing that makes multilevel modeling special is that it addresses specific violations of the 'assumptions' of OLS. Remember the assumptions? These are the conditions under which OLS regression works properly. All statistics have a set of assumptions under which they perform as they were intended.

One assumption is that the relationship between the independent and dependent variable is linear. Another is that the sample used in the analysis is representative of its population. Yet another is that the independent variables are measured without error. We more or less follow these assumptions in our day-to-day usage of OLS – and we should check that we are meeting some of these assumptions by doing regression diagnostics.

Dependence among observations

There are assumptions that relate to 'independence of observations' that we might think about less often. But it is this particular violation of the assumptions that multilevel modeling techniques are best suited to address. The assumption of independence means that cases in our data should be independent of one another, but if we have people clustered into groups, their group membership will likely make them similar to each other. Once people (Level 1) start having similar characteristics based on a group membership (Level 2), then the assumption of independence is violated. If you violate it, *you get incorrect estimations of the standard errors*. This isn't just a niggly pedantic point. If you violate the assumptions, you are more likely to wrongly achieve statistical significance and make conclusions that are simply incorrect.

Perhaps an overlooked common-sense fact is that even if you don't really care about group-level factors in your analysis (i.e. they aren't part of your hypotheses), ignoring them does not make the problem go away. This is easy to demonstrate.

Suppose we are interested in how gender and parental occupational status influence academic achievement. Table 1.1 presents results from an OLS regression of reading scores on gender and parental occupational status. Reading scores are standardized within the sample to a mean of 0 and a standard deviation of 1.

Gender is a dummy variable with 1 denoting males. Parental occupational status is a 64-category ordinal scale (the lowest value presenting low occupational status) that is treated as an interval-level variable here. The data come from the Australian sample from the 2006 Programme for International Student Assessment (PISA) organized by the Organisation for Economic Co-operation and Development (OECD, 2009). Data were obtained when all children were 15 and 16 years of age from schools in all eight states and territories of Australia.

Table 1.1   OLS regression of standardized reading scores on gender and parental occupational status ($N = 13,646$)

Variables	b	s.e.
Male	−0.381***	(0.016)
Parental occupational status	0.020***	(0.001)
Constant	−0.859***	(0.028)
R^2	0.141	

b – unstandardized regression coefficients; s.e. – standard errors

*$p < 0.05$, **$p < 0.01$, ***$p < 0.001$

Table 1.1 indicates that males (compared to females) have lower reading scores by 0.381 and that each unit increase in parental occupational status is associated with increases in reading scores of 0.020. These results are both statistically significant and have small standard errors. Our model has no group-level indicators, such as class or school. Just because we haven't included group-level indicators, it does not mean that our problems of dependence among observations and thus correlated errors do not exist.

First, we need to predict our residuals from the regression equation whose coefficients we have just identified. Remember that the residuals are the difference between our predicted reading score based on these characteristics and the actual reading score we see in the data. Next, we can test if the assumption is violated if we run an analysis of variance (ANOVA) of these residuals by a grouping factor. Our grouping factor here is the region of Australia – the state or territory. If the residuals are independent of the regions, that is great – that means our errors are randomly distributed. This is not, however, the case in our data as the ANOVA returns a result of $F = 55.8$, df = 7, $p < 0.001$.

It might be helpful to think of it this way: we have several thousand students within the eight different regions in Australia. The ANOVA tells us that our individual-level results violate that assumption of uncorrelated errors because we find that the ANOVA gives statistically significant results. Table 1.2 shows the mean reading scores by region.

Table 1.2 Means and standard deviations of standardized reading scores by region

Region	Mean	Std dev.
Australian Capital Territory	0.269	0.946
New South Wales	0.090	0.980
Victoria	−0.610	1.280
Queensland	−0.039	0.997
South Australia	0.070	0.915
Western Australia	−0.148	1.005
Tasmania	−0.075	0.937
Northern Territory	0.140	0.954

We could assume that a 'fix' to this would be to add dummy variables to the model that represent the different regions. We add dummy variables to a model so that we can include nominal-level variables in our estimation. As the regions are nominal, we can then add them as a set of dummies with an omitted reference category. Table 1.3 shows how many students are in each region in Australia, while Table 1.4 displays the regression results for the model including the region dummy variables.

Table 1.3 Australian students by region

Region	N	%
Australian Capital Territory	954	7.0
New South Wales	3,270	24.0
Victoria	702	5.1
Queensland	2,322	17.0
South Australia	1,548	11.3
Western Australia	1,221	9.0
Tasmania	2,183	16.0
Northern Territory	1,446	10.6
Total	13,646	100

As you can see in Table 1.3, students from this sample are dispersed among the eight regions of Australia. From reviewing the literature, we may have reason to believe that regional effects are determinants of academic achievement in Australia, as they are in many other countries around the world. For example, educational policies and resources are governed at the state level in Australia, and those regions containing the largest cities may offer the best resources for students (Australian Government, no date).

In Table 1.4 the gender and parental occupational status variables are the same as in Table 1.1. The region variable is entered as seven dummy variables, with the Australian Capital Territory as the omitted category.

Table 1.4 OLS regression of standardized reading scores on region, gender and parental occupational status ($N = 13,646$)

Variables	b	s.e.
Region [a]		
New South Wales	−0.075*	(0.034)
Victoria	−0.747***	(0.046)
Queensland	−0.168***	(0.035)
South Australia	−0.089*	(0.038)
Western Australia	−0.222***	(0.040)
Tasmania	−0.189***	(0.036)
Northern Territory	−0.026	(0.038)
Male	−0.377***	(0.016)
Parental occupational status	0.019***	(0.000)
Constant	−0.687***	(0.042)
R^2	0.165	

[a] Reference category is Australian Capital Territory

b – unstandardized regression coefficients; s.e. – standard errors

*$p < 0.05$, **$p < 0.01$, ***$p < 0.001$

Now is a good time to review the interpretation of coefficients as this will be important for understanding multilevel model outputs as well. The unstandardized coefficients (all in their own units of measurement) in Table 1.4 would be interpreted as:

- Compared to being in Australian Capital Territory, being in New South Wales is associated with a decrease in standardized reading scores by 0.075, controlling for the other variables in the model.
- Compared to being in Australian Capital Territory, being in Victoria is associated with a decrease in reading scores by 0.747, controlling for the other variables in the model.
- Compared to being in Australian Capital Territory, being in Queensland is associated with a 0.168 decrease in reading scores, controlling for the other variables in the model.
- Compared to being in Australian Capital Territory, being in South Australia is associated with a 0.089 decrease in reading scores, controlling for the other variables in the model.
- Compared to being in Australian Capital Territory, being in Western Australia is associated with a 0.222 decrease in reading scores, controlling for the other variables in the model.
- Compared to being in Australian Capital Territory, being in Tasmania is associated with a 0.189 decrease in reading scores, controlling for the other variables in the model.

- The coefficient for the Northern Territory is not statistically significant.
- Compared to being a female, being a male is associated with a 0.377 decrease in reading scores, controlling for the other variables in the model.
- Each unit increase in the parental occupational status scale is associated with a 0.019 increase in reading scores, controlling for all the other variables in the model.
- The constant is the reading score of –0.687, which is the value when all the independent variables have a value of zero. In this case it would be a female living in Australian Capital Territory whose parents have an occupational status score of zero – which isn't possible with this particular occupational status measure.

Clearly there are 'region' effects here – the Australian Capital Territory (ACT) seems to be the best place for reading scores. The problem with this type of analysis is that children are nested within the regions. Furthermore, they are nested within schools, and even classrooms. While our analysis is at the individual level, the observations aren't completely independent in the sense that there are eight regions into which pupils are divided.

An OLS regression assumes that the coefficients presented in the table are 'independent' of the effects of the other variables in the model, but, in this type of model, this assumption is false. If we believe that regions – an overarching structure – affect students differentially, the effect that regions have on reading scores is not independent of the effects of the other variables in the model. The structures themselves share similarities that we cannot observe but, nevertheless, influence our results. It is probably the case that parents' occupational status and the gender of the student are also differentially associated with reading scores, depending on the region in which the student attends school. The OLS assumption of independent residuals is probably violated as well. It is likely that the reading scores within each region may not be independent, and this could lead to residuals that are not independent within regions. Thus, the residuals are correlated with our variables that define structure. We need statistical techniques that can handle this kind of data structure. OLS is not designed to do this.

Group estimates

Maybe at this point you are wondering, 'If groups are so important, maybe I should just focus on group-level effects.' You may think that a possible solution to these problems is just to conduct analyses at the group level. In other words, to avoid the problem of giving group characteristics to individuals, just aggregate the data set so that we focus on groups, rather than individuals. In addition to having far less detailed models and violating some theoretical arguments (i.e. supposing your hypotheses are actually about individual- and group-level processes), a common problem with this approach is usually the sample size. If we have data on 13,000 students in eight different regions, aggregating the data to the group level would leave us with just eight cases (i.e. one row of data representing one region). To conduct any sort of meaningful multivariate analysis, a much larger sample size is required. As we mentioned earlier in this chapter, focusing on group estimates may lead to an error in logic known as the ecological fallacy where group characteristics are used to generalize to individuals. Thus, there are several reasons not to rely on group estimates.

Varying effects across contexts

In OLS, there is an assumption that the effects of independent variables are the same across contexts. For example, the effect of gender on school achievement is the same for everyone regardless of the region in which they go to school.

Table 1.5 OLS regressions of standardized reading scores on gender and parental occupational status, by region

	ACT	NSW	VIC	QLD
Male	-0.246***	-0.446***	-0.408***	-0.395***
	(0.056)	(0.031)	(0.093)	(0.039)
POS	0.024***	0.019***	0.020***	0.019***
	(0.002)	(0.001)	(0.003)	(0.001)
Constant	-1.047***	-0.720***	-1.436***	-0.840***
	(0.122)	(0.056)	(0.179)	(0.067)
N	954	3,270	702	2,322
	SA	WA	TAS	NT
Male	-0.410***	-0.386***	-0.336***	-0.279***
	(0.043)	(0.052)	(0.038)	(0.047)
POS	0.017***	0.024***	0.017***	0.018***
	(0.001)	(0.002)	(0.001)	(0.001)
Constant	-0.653***	-1.132***	-0.801***	-0.666***
	(0.076)	(0.084)	(0.065)	(0.086)
N	1,548	1,221	2,183	1,446

Unstandardized regression coefficients. Standard errors in parentheses. POS – parental occupational status; ACT – Australian Capital Territory; NSW – New South Wales; VIC – Victoria; QLD – Queensland; SA – South Australia; WA – Western Australia; TAS – Tasmania; NT – Northern Territory.

$^*p < 0.05, ^{**}p < 0.01, ^{***}p < 0.001$

We have many reasons to suspect that the effects of individual characteristics vary across contexts – that their impacts are not the same for everyone, regardless of group membership. We may find that the effects of gender are more pronounced in particular regions, for example. The context may influence the impact of gender on school achievement – for example, some regions may have an official policy around raising the science achievement of girls or the reading achievement of boys (as they are typical problems in the school achievement literature). Multilevel models allow regression effects (coefficients) to vary across different contexts (in this case, region) while OLS does not.

We might now think that one possible solution would be to run separate individual-level OLS regressions for each group. Table 1.5 displays the results of such an exercise.

This may seem to solve the problem of examining how group differences affect the impact of independent variables on the outcome of interest. You can see, for example, that the effect of being male ranges from being –0.446 in New South Wales (NSW) to –0.246 in the Australian Capital Territory (ACT). Likewise with parental occupational status (POS), the coefficients range from 0.017 in South Australia (SA) to 0.024 in the ACT and Western Australia (WA). These results do not tell us if the values are statistically significantly different from each other, without further calculations, and they also do not tell us anything about group properties which may influence or interact with individual-level outcomes. In addition to being a poor specification, this technique can get unwieldy if you have a large number of groups. Here we have only eight and the presentation of results is already rather difficult.

Another possible solution in OLS to effects varying across contexts might be to run interaction terms. You probably learned in your statistics training about interaction effects or moderating effects. If we thought that an independent variable affects an outcome differentially based upon the value of another independent variable, we could test this by using interaction terms. Based on the criticism above of running separate regressions for each group, a reasonable solution may seem to be to create interaction terms between the region and the other independent variables. We were making a similar argument earlier when we suggested that gender might impact on student achievement depending on region. We create the interaction terms by multiplying gender by region and parental occupational status by region (gender * regions; parental occupational status * regions) and we add them to the OLS regression as a set of new independent variables. If the interaction terms are statistically significant, it means that there is evidence that the effect of gender on reading and/or parental occupational status is contingent on the regions in which students go to school. The results of this estimation are presented in Table 1.6.

Table 1.6 OLS regression of standardized reading scores on gender, region, parental occupational status and interactions (*N* = 13,646)

Variables	b	s.e.
Male	−0.246***	(0.059)
Region [a]		
New South Wales	0.326*	(0.139)
Victoria	−0.389*	(0.184)
Queensland	0.207	(0.143)
South Australia	0.394**	(0.151)
Western Australia	−0.085	(0.153)
Tasmania	0.246	(0.144)
Northern Territory	0.380*	(0.154)
Male * New South Wales	−0.200**	(0.067)
Male * Victoria	−0.162	(0.091)
Male * Queensland	−0.149*	(0.070)
Male * South Australia	−0.164*	(0.075)
Male * Western Australia	−0.140	(0.079)
Male * Tasmania	−0.090	(0.071)
Male * Northern Territory	−0.033	(0.076)
Parental occupational status (POS)	0.024***	(0.002)
POS * New South Wales	−0.005*	(0.002)
POS * Victoria	−0.005	(0.003)
POS * Queensland	−0.005*	(0.002)
POS * South Australia	−0.007**	(0.002)
POS * Western Australia	0.000	(0.003)
POS * Tasmania	−0.007**	(0.002)
POS * Northern Territory	−0.007**	(0.003)
Constant	−1.047	(0.127)
R^2	0.167	

[a] Australian Capital Territory is the reference category

b – unstandardized regression coefficients; s.e. – standard errors

*$p < 0.05$, **$p < 0.01$, ***$p < 0.001$

The results in Table 1.6 suggest that region impacts on how individual characteristics affect the dependent variable. There are many statistically significant interactions. To really unravel what they mean, we have to look at them along with the main effects of the variables in the model. We can graph the main effects with the interaction effects and demonstrate the overall effect. We're not going to do that here, but we do address graphing interaction effects later. The main point from these results is that there are significant interactions. Perhaps we have just solved the problem of group effects?

Unfortunately not. There are still problems with this model. While there is no shortage of examples of this type of analysis in published work, one major problem with this approach is the nature of the cross-level interaction term. Cross-level interaction terms refer to interaction terms which have variables at different levels of aggregation. In this case, we have interacted individual characteristics (Level 1) with group characteristics (Level 2). This approach is fraught with problems. Treating group-level variables as though they are properties of individuals may result in flawed parameter estimations and downwardly biased standard errors (Hox and Kreft, 1994), and so we are more likely to find significant results. This also results in problems in the calculation of degrees of freedom, which leads to flawed estimates of the standard errors and faulty results. The problems associated with degrees of freedom are explained in more detail below.

Degrees of freedom and statistical significance

Degrees of freedom are a problem when using OLS to model multilevel relationships. When we use OLS and simply add group-level variables, such as region in the example above, we create a model that assumes individual-level degrees of freedom. At this point you may well be wondering, 'What are degrees of freedom?' – fair enough. As the name implies, degrees of freedom are related to how many of the values in a formula are able to vary when the statistic is being calculated. Our example data contains 13,646 students in eight different regions. These students then have 13,646 individual pieces of data. We use this information to estimate statistical relationships. In general, each statistic that we need to estimate requires one degree of freedom – because it is no longer allowed to vary. Many equations contain the mean, for example. Once we calculate a group mean, it is no longer able to vary. Again, once we calculate a standard deviation, we lose another degree of freedom. In the our examples above, degrees of freedom are determined from individual data, but if we have group characteristics in this individual-level data set, OLS calculates the degrees of freedom as though they are simply related to characteristics of individuals. In terms of group characteristics, the degrees of freedom should be based on the number of regions (8) rather than the number of pupils (13,646). The numbers – 13,646 versus 8 – are obviously very different. Degrees of freedom are integral in calculating tests of statistical significance. The resulting error from using the wrong degrees of freedom in OLS calculations is that it increases our likelihood of rejecting the null hypothesis when we should not. In other words, we are more likely to get statistically significant results – when we shouldn't – if we use the individual level degrees of freedom instead of the group level degrees of freedom.

Table 1.7   Summary of OLS 'workarounds' and their associated problems

Proposed solution	Unresolved problem
Ignore group-level characteristics, focusing solely on individual attributes.	Misspecification of model if group-level variables are indeed important predictors of outcome of interest. Also, if outcomes for individuals are correlated by group membership, even if groups are not considered in the model, there is still a problem with independence of observations (i.e. ignoring it doesn't make it go away). Possibility of committing atomistic fallacy.
Focus only on group variation and aggregate data to group level.	Ignores role of individual attributes in explaining outcome. May have problems if there are small numbers of groups. Possibility of committing ecological fallacy.
Separate regression estimations for each group.	Fails to address how the group properties may influence or interact with individual-level outcomes.
	There may be a problem if there are many groups and if there are small groups.
Create dummy variables for groups and create interaction terms between group dummies and individual-level characteristics.	Same problems as fitting regression estimates for each group separately. Degrees of freedom for cross-level interaction terms problematic. As well, the groups are treated as though they are unrelated although they may have things in common if they are drawn from a larger population of groups.

Adapted from Diez-Roux (2000: 173)

Table 1.7 summarizes the OLS 'workarounds' discussed above and their associated problems. The overarching problem is that when you use OLS models on data better suited to multilevel techniques you are very likely to *underestimate standard errors* and therefore increase the likelihood of results being statistically significant, possibly rejecting a null hypothesis when you should not. In other words, you are more likely to make a Type I error. If you correctly model your multilevel data then your results will be more accurately specified, more elegant, and more convincing, and your statistical techniques will match your conceptual model.

Multilevel modeling, in general and specific aspects of it, has also come in for some criticism. As with many debates of this kind, there is unlikely to be a firm and final conclusion, but we do advocate that users of any technique are aware of the criticisms and current debates. So we suggest that you start with this series of papers: Gorard (2003a, 2003b, 2007) and Plewis and Fielding (2003).

SOFTWARE

In the main text of this book we use Stata 13 software. We assume that the reader is familiar with the Stata software program as we do not see this book as an introduction to Stata – see Pevalin and Robson (2009) for such a treatment for an earlier version of Stata. At the time of writing, Stata 13 was the current version, with some changes for the main commands used in this book from version 12 – namely, the change from the **xtmixed** to the **mixed** command. While this changed little of the output and only a few options that are now defaults, it did impact on some of the other commands written by others for use after **xtmixed**. So, at times we use the Stata 12 command **xtmixed,** which works perfectly well in Stata 13 but is not officially supported. Stata 14 was released while this book was in production and we have checked that the do-files run in version 14.

As you may have gathered from the previous paragraph we use **this font** for Stata commands in the text. We italicize *variable names* in the text when we use them on their own, not part of a Stata command, but we also use italics to emphasize some points – we hope that the difference is clear.

In the text we use the **///** symbol to break a Stata command over more than one line. For example:

```
mixed z_read cen_pos || schoolid: cen_pos, ///
      stddev covariance(unstructured) nolog
```

This is only done to keep the commands on the book page. In the do-files (available on the accompanying webpage at the URL given below), the command can run on much further to the right. We have tried to keep what you see on these pages and what you see in the do-file the same, but if you come across a command without the **///** in the do-file then don't worry about it.

At a number of points we include the Stata commands that we used to perform certain tasks, including data manipulation and variable creation. As with all software, there are a number of ways to get what you want, some elegant and some a bit cumbersome. Our rationale is to start by using commands that are easy to follow and then move on to some of the more integrated features in Stata. In doing so, we hope that what we are doing is more transparent. If your programming skills are more advanced than those we demonstrate then we're sure you'll be able to think of more elegant ways of coding in no time at all. In the do-files to accompany this book we sometimes include alternate ways of programming to illustrate the versatility of the software.

There is a webpage to accompany this book at https://study.sagepub.com/robson. On this webpage you will find do-files and data files for each of the chapters so you can run through them in Stata and amend them for your own use. You will also find the chapter commands in R, with some explanation how to use R for these multilevel models. The webpage will be very much a 'live' document with links to helpful sites and other resources. In the list of references we have noted which papers are 'open access'. Links to these papers are also on the webpage.

We have chosen to use Stata in this book for two reasons. First, we wanted to use a general-purpose statistical software package rather than a specialized multilevel package. If readers already have experience of Stata then we avoid the challenges of getting to grips with new software. Even so, we could have then chosen from Stata, SPSS, SAS, and R. Which brings us to the second reason: Stata is our preferred software. It's what we use. It's what we teach with. And it's what we have previously written about. We chose to produce matching R code because R is a general-purpose software package and it is freely available. We both have previous experience with SPSS, and so SPSS code for the examples may well materialize on the webpage. For those who can't wait, Albright and Marinova (2010) have produced a succinct primer for 'translating' multilevel models across SPSS, Stata, SAS, and R which is available at: http://www.indiana.edu/~statmath/stat/all/hlm/hlm.pdf

If your preference is for a specialized multilevel package then we strongly recommend that you start with the resources available from the Centre for Multilevel Modelling at the University of Bristol and their MLWiN software (free to academics) at: http://www.bristol.ac.uk/cmm/

HOW THIS BOOK IS ORGANIZED

This book is organized into four chapters. This introductory chapter is followed by a chapter on random intercept models. This starts with a very quick review of OLS single-level regression to get us all on the same page and then gradually introduces random intercept models, how these differ from single-level OLS regression, and how to explore the nesting structures in your data. There are a few points where we touch on ongoing debates in the multilevel modeling community. Not everything is clear-cut, but we try to steer you through these without getting stuck in the technicalities (and probably a never-ending debate!). We continue with developing random intercept models, building up slowly with plenty of illustrations and example commands and output which is explained. We try not to throw too many equations at you, but when we do, we explain them in words. We look at adding independent variables at different levels, interaction effects, model fit statistics, and model diagnostics. Throughout, we try to follow the same example, but at times we have to deviate from that to demonstrate some points.

Chapter 3 is about random coefficient models, sometimes called random slope models. These are more complicated models than random intercepts, and we recommend that you read through Chapter 2 before starting out on this chapter. We pretty much follow the same topics as with random intercept models.

Of course there will be more detail in the chapters, but to conclude this introductory chapter we will simply remark that random intercept models look like this, with parallel regression lines for each group:

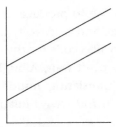

And random coefficient models look like this, with group regression lines which are not parallel to each other:

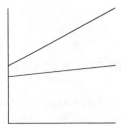

In Chapter 4 we turn to how to present results from these models and how to use Stata to produce publication-quality tables.

CHAPTER 1 TAKEAWAY POINTS

- If your data contain nested levels, you should probably use multilevel modeling.
- Ignoring the levels in your data and simply using ordinary regression techniques requires you to violate important assumptions in regression theory.
- Multilevel modeling techniques allow you to better specify your model and achieve more accurate results.
- This book only considers cross-sectional nested data.
- Your models should be theory-driven and not motivated by the perceived need to add arbitrary complexity to your estimations.

TWO

Random Intercept Models: When intercepts vary

CHAPTER CONTENTS

In this chapter we introduce random intercept models. These are models in which the intercept is allowed to vary for each Level 2 group; this is the random part of the model. For example, in Chapter 1 we looked at the reading scores of children in the eight regions of Australia. The Level 2 group is the region and, so, in a random intercept model we would have eight intercepts – one for each region. As there are as many intercepts as there are groups (there could be hundreds!), these cannot be represented as a single coefficient or value as in ordinary regression. Most statistical software output uses a measure of variation such as the standard deviation or variance to capture this information. If this variation is significantly different from zero then we know that a random intercept model is better than the single intercept model used in ordinary regression. These models have a fixed part too. This is where the independent variables go. As with ordinary regression, a coefficient is calculated for each independent variable along with statistics to judge its significance such as its standard error, z value, p value, and confidence intervals. These fixed coefficients are interpreted in the same way as in ordinary regression.

We start with a quick review of ordinary regression before easing into the random intercept models. We offer a step-by-step approach to exploring the data and specifying models with main effects and interaction effects. We take time to go through the Stata output in detail and focus on the interpretation of the results. Centring variables is a common practice in multilevel modeling and we spend time demonstrating and discussing the different ways to do this and how to interpret results from centred variables. We finish the chapter by discussing measures of model fit, weighting and diagnostics.

A REVIEW OF SINGLE-LEVEL REGRESSION

In our quick review of single-level regression, we will continue to use our example that investigates the effects of parental occupational status on standardized reading scores. In ordinary regression the model is usually described with the following equation:

$$y_i = \beta_0 + \beta_1 X_i + e_i$$

In plain language:

- y_i is the dependent (or outcome) variable of interest;
- β_0 is the intercept, or the point on the Y-axis (vertical axis) that the line meets;
- β_1 is the coefficient (or slope) that determines the trajectory of the line, while X_i is the observed value of our independent variable;
- e_i is the error, or the residual, which is the difference between our estimated and observed values;
- the subscript i refers to the observation number.

Turning to our example of reading scores as our outcome, let's start by thinking of $y_i = \beta_0 + \beta_1 X_i + e_i$ as

$$z\_read_i = \beta_0 + \beta_1 POS_i + e_i$$

where we predict standardized reading scores (z_read) based on a single independent variable, parental occupational status (pos).

Therefore in this model, β_1 represents the average effect of parental occupational status on standardized reading scores and represents the intercept. The value of is the predicted value of standardized reading scores when parental occupational status is zero. Depending on how parental occupational status was measured, this may or may not make sense.

If we put these variables into simple regression in Stata, we get the results shown in Table 2.1.

```
regress z_read pos
```

Table 2.1 Stata output for the regression of standardized reading scores on parental occupational status

```
      Source |       SS           df       MS          Number of obs   =    13646
-------------+----------------------------------        F(1, 13644)     =  1602.03
       Model |  1433.79854         1   1433.79854       Prob > F        =   0.0000
    Residual |  12211.2015     13644   .894986915       R-squared       =   0.1051
-------------+----------------------------------        Adj R-squared   =   0.1050
       Total |              13645 13645           1     Root MSE        =   .94604

------------------------------------------------------------------------------
      z_read |      Coef.   Std. Err.      t    P>|t|     [95% Conf. Interval]
-------------+----------------------------------------------------------------
         pos |     .019765   .0004938   40.03   0.000     .0187971    .020733
       _cons |   -1.047046   .0273844  -38.24   0.000    -1.100724   -.993369
------------------------------------------------------------------------------
```

Translating these results from Table 2.1 gives us:

(predicted) standardized reading score = −1.047 + 0.020(parental occupational status)

The value at which the regression line crosses the Y-axis is −1.047 when parental occupational status is zero. However, as the variable *pos* ranges from 16 to 90, the value of zero is not meaningful. If the *pos* variable was rescaled so that zero was a meaningful value (in this case, the mean), then the Y intercept represents a meaningful value. If you look at the descriptive statistics for the variable *pos* you will find that it starts at 16; however, while this is more meaningful than zero, as it is an actual value in the sample, it may be more meaningful and easier to interpret if the variable was centred on the mean. We will return to the idea of centring shortly. To keep with good pedagogy and to make the further examples more meaningful, it helps if we centre the *pos* variable. Centring may also be called grand mean centring and refers to rescaling a variable wherein we create a new variable by subtracting the mean of the variable from the its observed scores. This results in a centred variable that has a mean of zero and keeps its original units of measurement. However, centring does not have to be around a sample characteristic, such as the sample mean. It can also be around a typical value or a known population average (something we return to in Chapter 3).

This is easily accomplished by subtracting the mean *pos* score from the observed scores. Below we first summarize the variable *pos*. Stata stores the results of the calculations for use in other commands. The command **return list** shows us what has been stored. The values are called scalars and use the notation r(X).

```
su pos

return list

scalars:
                r(N)    = 13646

            r(sum_w)    = 13646

             r(mean)    = 52.97464458449362

              r(Var)    = 268.9793497270967

               r(sd)    = 16.40058992009424

              r(min)    = 16

              r(max)    = 90

              r(sum)    = 722892
```

The scalars can then be used in subsequent commands, providing they have not been overwritten. In this case we use the scalar for the mean, r(mean):

```
gen cen_pos = pos - r(mean)
```

Rerunning the regression with the centred variable we get the results in Table 2.2.

```
regress z_read cen_pos
```

Table 2.2   Stata output for the regression of standardized reading scores on parental occupational status (centred variable)

Source		SS	df	MS		Number of obs	=	13646
						F(1, 13644)	=	1602.03
Model		1433.79854	1	1433.79854		Prob > F	=	0.0000
Residual		12211.2015	13644	.894986915		R-squared	=	0.1051
						Adj R-squared	=	0.1050
Total		13645	13645	1		Root MSE	=	.94604

z_read		Coef.	Std. Err.	t	P>|t|	[95% Conf. Interval]	
cen_pos		.019765	.0004938	40.03	0.000	.0187971	.020733
_cons		-1.72e-08	.0080985	-0.00	1.000	-.0158742	.0158742

The value of the coefficient for the centred variable (*cen_pos*) is the same as for the original variable (*pos*) in Table 2.1, but the constant is now zero. So, when parental occupational status is at its mean value (which is now zero – e–08 means 10 to the power of –8, which puts seven zeros in front of the 1.72!), the predicted standardized reading score is zero – the mean of the variable *z_read*. This is not surprising as a regression line always passes through the point on a graph located by the mean of x (the independent variable) and the mean of y (the dependent variable) and in this case both of our variables have a mean of zero: parental occupational status because we centred it about its mean, and standardized reading scores because the reading scores were standardized within the sample.

If we graph the two variables in Figure 2.1, we can overlay the fitted line on the scatterplot of the independent and predicted values of the dependent variable:

```
twoway (scatter z_read cen_pos) || (lfit z_read cen_pos)
```

This is not a very good model. Most real-life models are far from perfect, and this is a great example of an imperfect model. Note the scatter around the fitted line. The fitted line is the average – the line of least square deviations between the observed and predicted values, to be precise. We can see that there is a lot of 'noise' – or a lot of 'lack of fit'.

This simple model forces the equation to have one intercept for all observations. We may have good reason to suspect that it makes more sense to hypothesize that the groups may have different intercepts. It may be the case, for example, that if we

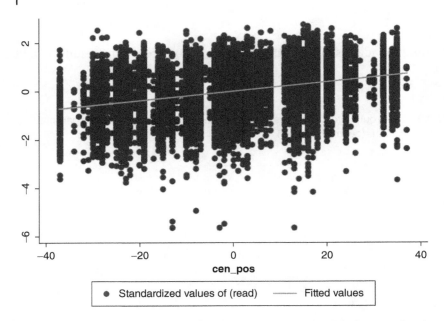

Figure 2.1 Scatterplot of standardized reading scores by centred parental occupational status, with regression line

allowed each *region* to have its own intercept, we would end up with a better, more meaningful model as it would be plausible that the educational policies, practices and priorities vary from region to region and, so, may result in different levels of reading ability. In addition, Chapter 1 has provided several reasons for why we would want to pursue other techniques. First and foremost, we are violating the assumptions by working with nested data using traditional regression techniques.

NESTING STRUCTURES IN OUR DATA

In Chapter 1, we showed a number of models that included region as a determining factor in predicting standardized reading scores. We will continue this example here, using region as a nesting structure. Recall that there are eight regions within the data. We work with the hypothesis that region is an important factor in predicting reading scores because children who are schooled in the same region will have similar characteristics which influence how well they do on the standardized reading test.

How many groups?

You are probably wondering if eight groups (in this case, regions) are 'enough' to do multilevel modeling. This is an interesting question because there is debate among the experts in the field about this very topic.

Richter (2006) argues that we need at least 30 groups comprised of at least 30 observations to have reliable estimates, referring to this as the 30–30 rule, and Hox (1998) advocates a 50–20 rule, that is, at least 50 groups comprised of at least 20 observations at Level 1; both rules well above the eight regions we are considering using here. Conversely, Nezlek (2008) and Gelman (2006) suggest that 10, or fewer, groups may be adequate but you may run into estimation problems. Maas and Hox (2005) and Bryan and Jenkins (2013) have found that a small number of groups leads to problems with the estimates of Level 2 variables. As Bryan and Jenkins (2013: 2) say: 'The intuition is straightforward: in general, desirable properties of regression model parameter estimates such as consistency and efficiency are contingent on sample sizes being "large".' In most circumstances we would not advocate using multilevel modeling with only eight groups, but for our purposes here the small number of groups helps the clarity of our example. We soon move onto more a realistic number of groups.

So what do you do if your group numbers are small? Rather than abandoning the multilevel structure and assuming the group variance is zero, it is probably a better idea to use multilevel modeling. The only other alternative is to ignore group variance, which would be a bigger mistake, because much of the content of this book thus far has been focused on arguing the importance of properly specifying the theoretical and empirical relationships in our data. Thus, if your group numbers are small – less than 20, for example – be aware that there may be some estimation biases in your Level 2 variables and, so, be very cautious with conclusions from the Level 2 results.

Sample sizes within groupings

What about the number of Level 1 units in your Level 2 groupings? Mok (1995) and Snijders (2005) have both determined that the number of groups is more important in determining statistical power than the number of Level 1 units within the groups. As we are most likely to be using secondary data, we are not in a position to decide how the units are measured, but in general current opinion suggests the more Level 2 units the better, and that the sample sizes within the groups are of lesser importance than the number of groups themselves.

What sorts of variables can be levels?

In our current example, we use region as a Level 2 variable. Often, whether a variable is a 'level' is quite straightforward: regions, schools, households, neighbourhoods are all common Level 2 variables. These decisions are not entirely straightforward. Duncan et al. (1998) suggest that a level must have several different groups, as mentioned above. If your variable only has two or three groups, then it is probably not a level. As discussed earlier, the minimum number of groups required is a debated topic but, generally, the more, the better. Variables that can be treated as levels are always nominal in terms of their level of measurement. Their categories may have numbers assigned to them, but the numbers are just placeholders and have no special quantitative value and they cannot be ordered in any meaningful way.

Another factor in determining whether a variable is a level is its 'exchangeability', which is a somewhat complex concept. In a very general sense, this means that individual groups within the level are not particularly special – that they are not substantively interesting – and that they all come from the same population. In other words, we are not interested in any one particular region over any other. We should not have any reason to believe that any one of our groups should predict a different result – if we do have reasons to predict this, then the variable should probably be considered as an explanatory variable rather than a level. For example, if we wanted to investigate private schools compared to public schools, we would enter this characteristic as an independent variable at Level 2 as it is a characteristic of the schools rather than the students.

Therefore, we must ask ourselves whether region is a level. It has a small number of groups and we may have reason to suspect that students in the richer and more densely populated parts of Australia do better in school. Although we highlight this potential problem, we will continue to consider region as a level for the first part of the forthcoming examples. As in all data analysis, we must make decisions, and real-life data analysis problems are often not as neat, perfect, and straightforward as the examples usually provided in textbooks!

GETTING STARTING WITH RANDOM INTERCEPT MODELS

The first step in multilevel modeling is to run a null or empty model with no covariates. A two-level model without independent variables, the null model, can be expressed as:

$$y_{ij} = \beta_o + \mu_j + e_{ij}$$

We can dissect this formula into its components with the assistance of the illustration in Figure 2.2. Other authors may use different notation, but the formula is essentially the same. Subscripts can be other letters, not just i and j, and characters may be something other than μ (e.g. ξ) but this does not change the meaning of the equation.

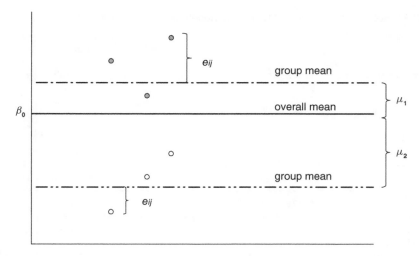

Figure 2.2 Graphical representation of a null model with two groups

- y_{ij} is the dependent variable. The subscripts i and j refer to the two levels that are in the model. i refers to the Level 1 units, in our example students, and j to the groups at Level 2, in our current example the regions.
- β_0 is the overall mean of the dependent variable across all groups.
- μ_1 is the random intercept for Group 1 and μ_2 is the random intercept for Group 2.
- e_{ij} is the residual, which is the difference between the ith student's standardized reading score and their group's mean.

In Stata, we can run a null multilevel model using the command **xtreg**. We run a null model with no independent variables in order to first explore whether there is any evidence of random intercepts for the grouping Level 2 variable (*region*). The null model gives us the first bit of information we need in the process of deciding whether or not to pursue multilevel modeling.

The **xtreg** command is straightforward, with our dependent variable added first, then 'i' denoting our grouping variable (in this case, *region*). The option **mle** refers to the algorithm that we are using, which in this case is maximum likelihood. Other algorithms are used, but maximum likelihood is by far the most popular and accepted one in the field at the time of writing. We use the old option **i(region)** to denote the group variable. This still works in version 13 and saves you the need to declare the structure of the data using the **xtset** command.

```
xtreg z_read, i(region) mle
```

Table 2.3 Stata output for region groups null model using `xtreg` command

```
Random-effects ML regression                    Number of obs       =     13646
Group variable: region                          Number of groups    =         8

Random effects u_i ~ Gaussian                   Obs per group: min  =       702
                                                               avg  =    1705.8
                                                               max  =      3270

                                                Wald chi2(0)        =      0.00
Log likelihood  = -19163.244                    Prob > chi2         =         .

-------------------------------------------------------------------------------
      z_read |      Coef.   Std. Err.      z    P>|z|     [95% Conf. Interval]
-------------+-----------------------------------------------------------------
       _cons |   -.037039   .0872234   -0.42   0.671    -.2079938    .1339157
-------------+-----------------------------------------------------------------
    /sigma_u |   .2452823   .0625322                     .1488199    .4042699
    /sigma_e |   .9841617   .0059591                     .9725511    .9959108
         rho |   .0584827   .0280836                      .020721    .1364964
-------------------------------------------------------------------------------
Likelihood-ratio test of sigma_u=0: chibar2(01)= 398.18 Prob >= chibar2 = 0.000
```

The output in Table 2.3 tells us many things. Starting at the top:

- There are 13,646 observations and the grouping variable, *region*, has 8 groups.
- There are a minimum of 702 observations per group to a maximum of 3,270 observations per group, with the average around 1,706.
- The log likelihood is –19,163.244, which isn't terribly important until we start comparing different models later in the chapter.
- _cons is the constant and is –0.037. This is the mean weighted standardized reading score for the total sample. As the reading scores were standardized, you may expect this to be zero, but the constant reported in the output weights the observations by region.
- sigma_u is the between-group standard deviation or the random intercept of the Level 2 (grouping) variable and is 0.245. It is important to examine this with the last line of the output, which shows the likelihood ratio test of sigma_u = 0.
- sigma_e is the within-group, or between-subject at Level 1, standard deviation and is 0.984.
- Around our estimates, 95% confidence intervals are also provided. We discuss these in more depth after Table 2.11.
- rho is called the intraclass correlation coefficient (ICC) and tells you how much of the variation in the dependent variable can be attributed to differences between your Level 2 variable (in this case, *region*). Here our ICC is 0.058. Therefore, 5.8% of the variation in standardized reading score can be attributed to differences between regions. The rest can be attributed to differences between individuals or other groupings.
- The likelihood ratio test of sigma_u = 0 is testing the null hypothesis that the standard deviation (sigma_ u) of the random intercept of the Level 2 grouping variable is equal to zero. The value of the likelihood ratio test here is 398.18 with a corresponding *p* value stated as 0.000, and meaning less than 0.0005, which gives us strong evidence in favour of rejecting the null hypothesis. In other words, there is evidence that using a random intercept model helps

explain the variance in standardized reading scores, even in the absence of any covariates (or independent variables). Note that the output uses chibar2, which is not exactly the same as a chi-square statistic. This means the p value is cautious; in other words, it overestimates the value of p. So, if the reported value is less than 0.05 then the accurate value will also be less than 0.05. Just be careful with values around 0.05. Another way of thinking about this is that it indicates that there is significant variation in the intercepts. If there wasn't significant variation then we could justifiably return to a regression model with a single intercept.

We can get similar results in Stata by using the command **mixed**. The notation for **mixed** has the 'fixed' part of the equation on the left-hand side and the 'random' part after the two vertical bars on the right-hand side. We find it useful to think of the vertical bars as a fence that separates the fixed from the random parts. Here, we don't have any fixed parts of the equation because we are running a null model with no covariates. On the right-hand side, we have a random intercept for *region*, which is followed by a colon. In Chapter 3 we will add 'random coefficients' after this colon, but for now we have nothing there, which indicates that there are no random coefficients to estimate. In Stata version 12 and higher, maximum likelihood is the default estimator for the **mixed** command. The **mixed** command produces variances as the default rather than standard deviations, so for comparability with the **xtreg** output we use the **stddev** option.

```
mixed z_read || region:, stddev
```

Table 2.4   Stata output for region groups null model using `mixed` command

```
Mixed-effects ML regression              Number of obs     =         13646
Group variable: region                   Number of groups  =             8

                                         Obs per group: min =           702
                                                        avg =        1705.8
                                                        max =          3270

                                         Wald chi2(0)      =             .
Log likelihood = -19163.244              Prob > chi2       =             .

------------------------------------------------------------------------------
   z_read |     Coef.    Std. Err.     z     P>|z|     [95% Conf. Interval]
----------+-------------------------------------------------------------------
    _cons | -.037039     .0872258   -0.42   0.671    -.2079985    .1339204
------------------------------------------------------------------------------


------------------------------------------------------------------------------
  Random-effects Parameters |   Estimate   Std. Err.    [95% Conf. Interval]
----------------------------+-------------------------------------------------
region: Identity            |
               sd(_cons) |   .2452824   .0625371    .1488141    .4042858
----------------------------+-------------------------------------------------
            sd(Residual) |   .9841617   .0059591    .9725511    .9959108
------------------------------------------------------------------------------
LR test vs. linear regression: chibar2(01) =  398.18 Prob >= chibar2 = 0.0000
```

If you compare the results of **mixed** (Table 2.4) and **xtreg**, you will see they are nearly identical, apart from details in how the results are presented. The constant is still called _cons, but the standard deviation of the Level 2 variable is now called sd(_cons) instead of sigma_u and the within-group between-student standard deviation is called sd(Residual) instead of sigma_e.

The last line of the output performs a similar test to the last line of the **xtreg** output except that this is framed as a test of this model versus a single-level regression model. Notice that the values are the same, so again this indicates that allowing the intercept to vary by group significantly improves the model.

You will notice that **mixed** does not automatically produce rho, the ICC. This can be annoying, but in version 12 and higher there is a Stata post-estimation command **estat icc** which will produce the ICC along with its standard error and 95% confidence interval. If you have an earlier version of Stata without the **estat icc** post-estimation command then there is a command called **xtm-rho** which can be easily downloaded by typing **ssc install xtmrho** (if not already installed). After running **mixed**, you get the ICC by typing **xtmrho**. The **xtmrho** command has other uses (see later in this chapter), so we recommend you install it.

```
estat icc

Intraclass correlation

------------------------------------------------------------------------------
        Level |       ICC     Std.     Err.          [95% Conf. Interval]
--------------+---------------------------------------------------------------
       region |   .0584828   .0280858        .0223466        .1444219
------------------------------------------------------------------------------
```

The random intercept model is a special case of 'random effects' models that can be estimated in Stata using the **xtreg** command. Typically, this command is used for panel regression techniques. Multilevel models can contain fixed and random parts in the intercept and coefficients, which cannot be accommodated by **xtreg**. Such models will be explored later in this book. There are also added features and options in the **mixed** command that are specifically orientated to multilevel modeling. However, **xtreg** is very convenient for exploring the simple elements of variance components and has the added advantage of running quite quickly. Sometimes the **mixed** command can take longer to run. So, in the exploratory stages of your research, you may prefer to use the **xtreg** command due to its faster computation speed.

WHAT DO OUR FINDINGS MEAN SO FAR?

Referring back to Figure 2.2, this shows that the model we just ran would produce a total sample intercept (for all groups) of –0.037 in our output (Tables 2.3 and 2.4). We would also have eight intercepts, one for each group. This can be quite a strange concept to try to imagine with actual data, so let us try to visualize this a bit more.

After the **mixed** command, we can use the **predict** commands in Stata to produce our group intercepts and standard errors:

```
predict u0, reffects
```

Here we are creating a variable called *u0* that is the predicted random intercepts (group-level residuals, or how much the group value deviates from the overall sample intercept: β_0 in Figure 2.2) from our model. We use the option **reffects** to tell Stata that we wish to have the values of our random intercepts. You will notice at the bottom of your variable list that there is now a variable labelled 'BLUP r.e. for _cons', which means best linear unbiased prediction random effect for constant.

```
predict u0se, reses
```

Next, we create a variable called *u0se* that is the standard errors of the random intercepts in our model, and for this we use the option **reses**. You will notice in your variable list that there is now a variable that is labelled 'BLUP r.e. std. for region _cons', which means best linear unbiased prediction random effect standard error for region constant.

Every individual student in the same region has the same values for the variables *u0* and *u0se*. In order to produce tables and graphs that are meaningful, we need just one observation per group. We can do this by creating a variable called *justone* using the **egen** command that selects one observation per group using the option **tag**:

```
egen justone=tag(region)
```

We will also create a variable that ranks the region effects from lowest to highest for easier interpretation and graphing. We first sort the random intercept values and then create a new variable *rank*, which is the running sum of *justone*:

```
sort u0
generate rank=sum(justone)
```

Now we can list these variables and examine the random intercept and its standard error for each group, and these are shown in Table 2.5.

```
sort rank

list region u0 u0se rank if justone==1
```

Table 2.5 Stata output of ranked random intercept values by region

```
+------------------------------------------+
| region         u0        u0se    rank |
|------------------------------------------|
|    VIC   -.5598428    .0367261       1 |
|     WA   -.1099004    .0279811       2 |
|    TAS   -.0379094    .0209867       3 |
|    QLD   -.0018652    .0203533       4 |
|     SA    .1062203    .0248848       5 |
|------------------------------------------|
|    NSW     .126686    .0171683       6 |
|     NT    .1754888    .0257382       7 |
|    ACT    .3011227    .0315979       8 |
+------------------------------------------+
```

Recall from the output of the null model (Table 2.4) that the intercept (_cons) for the total sample was −0.037, the weighted mean for the total sample. For each region, the variable $u0$ tells you how much to add or subtract from −0.037 to get the intercept for that particular region – the values for μ_1 or μ_2 in Figure 2.2. For example, the intercept for the Australian Capital Territory would be −0.037 + 0.301 = 0.264, which means its intercept is above the total sample intercept. But the intercept for Victoria would be below at −0.037 − 0.559 = −0.596. The rankings show you the order of the random intercepts in terms of rank, with number 1 being the furthest below the total sample intercept and number 8 the highest above.

The $u0se$ variable gives us the information we need to create confidence intervals around the estimates of the random intercepts, which we can display using the command:

```
serrbar u0 u0se rank if justone==1, scale(1.96) yline(0)
```

This command produces a standard error bar chart (Figure 2.3), with 95% confidence intervals (denoted by the **scale(1.96)** option). The option **yline(0)** places a horizontal line at zero. The values of $u0$, indicated by the solid circle, are relative to the total sample intercept, so the graphed values are deviations rather than actual values of the intercepts.

This plot is also known as a caterpillar plot. We only have eight data points, so it does not look much like a caterpillar, unfortunately.

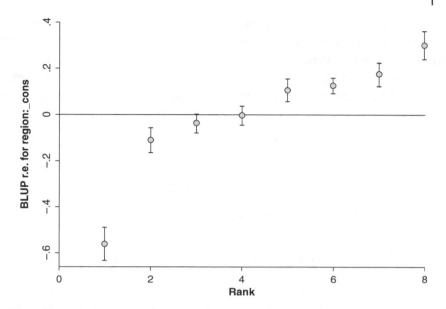

Figure 2.3 Standard error bar chart of relative random intercept values for region null model

Is this model a good one?

Now that we understand what our output means and how to visualize it, we need to decide if it is worth pursuing multilevel modeling. There are two statistics you can look to for guidance on this matter.

The first is the likelihood ratio test statistic reported at the bottom of the **xtreg** and **mixed** output (Tables 2.3 and 2.4). This test compares the null multilevel model with the null single-level (i.e. ordinary regression) model. If this test is statistically significant (i.e. the *p* value is less than 0.05), we can conclude that there is evidence of group effects in our data. In the output, the chi-square value is given as 398.18 with a *p* value stated as 0.000, indicating that there are significant group differences in our model.

The second is rho, the ICC. The ICC reported in this model was 0.058, meaning that 5.8% of the variance in the model was explained by differences between regions. Is this good? There is no objective cut-off for what constitutes a good ICC. The value of an ICC ranges from 0 to 1. If it is zero, or close to zero, there is no evidence of multilevel or nesting effects in your data. If it is 1, the nesting accounts for all the variance in your model. In other words, there is no variance at the individual level in your model – everyone is the same within their groups. Some experts have advocated a somewhat flexible guideline of around 0.10 as being indicative of a 'non-trivial' ICC (Lee, 2000), although multilevel models with ICCs less than 0.10 have been published. Nezlek (2008) warns of using a low

ICC value to justify not using multilevel modeling and advises concentrating on the data structure to decide whether or not to use multilevel modeling rather than seeing what the ICC brings.

So far it seems that the likelihood ratio test is suggesting that there are important group differences but the ICC value of 0.058 seems rather small. It might be a better idea to explore other nesting features in the data and see how they compare.

Calculating the ICC

Calculating the ICC is not particularly difficult:

$$\text{ICC} = \frac{\text{variance of random intercept}}{\text{variance of random intercept} + \text{within-subject variance}}$$

Remember that the variance is just the square of the standard deviation (or conversely, that the standard deviation is the square root of the variance). Stata 13 reports the variance as the default in the output.

```
mixed z_read || region:
```

Table 2.6 Stata output with variances for region groups null model using `mixed` command

```
Mixed-effects ML regression              Number of obs      =       13646
Group variable: region                   Number of groups   =           8

                                         Obs per group: min =         702
                                                        avg =      1705.8
                                                        max =        3270
                                         Wald chi2(0)       =           .
Log likelihood = -19163.244              Prob > chi2        =           .

------------------------------------------------------------------------------
    z_read |     Coef.   Std. Err.      z    P>|z|     [95% Conf. Interval]
-----------+------------------------------------------------------------------
     _cons |   -.037039   .0872258    -0.42   0.671    -.2079985    .1339204
------------------------------------------------------------------------------


------------------------------------------------------------------------------
  Random-effects Parameters |   Estimate   Std. Err.    [95% Conf. Interval]
-----------------------------+------------------------------------------------
region: Identity             |
                 var(_cons)  |   .0601634   .0306785    .0221456     .163447
-----------------------------+------------------------------------------------
              var(Residual)  |   .9685742   .0117294    .9458557    .9918383
------------------------------------------------------------------------------
LR test vs. linear regression: chibar2(01) =   398.18 Prob >= chibar2 = 0.0000
```

In the output in Table 2.6, the Level 2, between-region variance is estimated as 0.060 and the Level 1, between-student within-region variance is estimated as 0.969. Compare this output with that in Table 2.3. For example, in Table 2.3 the estimate for sd(_cons) was 0.245, and $0.245^2 = 0.060$, which is the var(_cons) estimate.

The estimates var(_cons) and var(Residual) are the two variance components. In other words, the variance is allocated to different parts of the model: between regions, and between students within regions. The total variance of the model is calculated by just adding them up: $0.060 + 0.969 = 1.029$.

Therefore, the ICC is $0.060/(0.060 + 0.969) = 0.058$.

CHANGING THE GROUPING TO SCHOOLS

We started exploring the regional effects on standardized reading scores. A cursory examination of the literature on school achievement would quickly reveal that schools are a major nesting variable of interest for many education researchers. Indeed, the vast majority of examples you will find online and elsewhere about multilevel modeling will use the example of nesting students within schools – it is a very common thing to do in multilevel modeling when using student data.

Before we move on to the practicalities of using schools as a grouping variable, we would just like to briefly note one point. It may have occurred to you that when we were using regions we had all the states and territories in Australia as groups. Therefore, the regions were fully observed and not a sample of regions in Australia. When we use schools as the grouping variable, we will be using a sample of schools drawn from a finite population – all schools in Australia. Similar situations of fully observed grouping variables could occur if, for example, you had data from all states in the USA, all hospitals in a county of England, or all countries that are part of the EU. These fully observed groups require slightly different thinking than when using samples of infinite populations, usually invoking the somewhat controversial idea of super-populations. For the purposes of this introductory book, this is all we are going to say on this point. If you are interested in further reading then we recommend starting with Ranstam (2009), a freely available primer, and the critical view from Gorard (2007).

The variable in the data that records the student's school identifier is *schoolid*. We could run a **tabulate** of this variable to get an idea of how many schools there are and how many students there are within each school, but this may not be practical if there are a large number of schools. Alternatively, we can create some simple variables in Stata that, when combined together, will give us useful summary information on the characteristics of *schoolid*:

```
egen schnum=count(schoolid), by(schoolid)

egen justone=tag(schoolid)

su schnum if justone==1
```

Table 2.7   Summary statistics to determine number of schools

Variable	Obs	Mean	Std. Dev.	Min	Max
schnum	356	38.33146	10.03275	3	57

We can tell from the output in Table 2.7 that there are 356 different schools with a minimum number of observations (students) of 3 and a maximum of 57. The average is around 38 students per school.

To evaluate if there are 'school effects' on standardized reading scores, the first step is to again run the null model. The notation is exactly the same as when we used the *region* variable, except that *schoolid* is in the parentheses in the **xtreg** command:

```
xtreg z_read, i(schoolid) mle
```

The output is in Table 2.8.

Table 2.8   Stata output for school groups null model using xtreg command

```
Random-effects ML regression           Number of obs      =       13646
Group variable: schoolid               Number of groups   =         356

Random effects u_i ~ Gaussian          Obs per group: min =           3
                                                      avg =        38.3
                                                      max =          57

                                       Wald chi2(0)       =        0.00
Log likelihood    = -18024.825         Prob > chi2        =           .

-------------------------------------------------------------------------
      z_read |     Coef.   Std. Err.     z    P>|z|   [95% Conf. Interval]
-------------+-----------------------------------------------------------
       _cons |  -.0365929  .0286407   -1.28   0.201   -.0927276   .0195418
-------------+-----------------------------------------------------------
   /sigma_u  |   .5179856  .0220386                    .4765427   .5630325
   /sigma_e  |   .8761847  .0053806                    .8657021   .8867941
        rho  |   .2589834  .0165613                    .2276053   .2924612
-------------------------------------------------------------------------
Likelihood-ratio test of sigma_u=0: chibar2(01)= 2675.02 Prob >= chibar2 = 0.000
```

As in the previous example, we can also get to these results in Stata by using the command **mixed** but producing variances instead of standard deviations (Table 2.9):

```
mixed z_read || schoolid:
```

Table 2.9 Stata output for school groups null model using `mixed` command

```
Mixed-effects ML regression              Number of obs      =      13646
Group variable: schoolid                 Number of groups   =        356

                                         Obs per group: min =          3
                                                        avg =       38.3
                                                        max =         57

                                         Wald chi2(0)       =          .
Log likelihood = -18024.825              Prob > chi2        =          .

------------------------------------------------------------------------------
      z_read|      Coef.   Std. Err.     z    P>|z|     [95% Conf. Interval]
-------------+----------------------------------------------------------------
       _cons|   -.0365929   .0286371   -1.28  0.201    -.0927206    .0195348
------------------------------------------------------------------------------


------------------------------------------------------------------------------
 Random-effects Parameters   |   Estimate   Std. Err.   [95% Conf. Interval]
-----------------------------+------------------------------------------------
schoolid: Identity           |
                 var(_cons)  |   .2683094   .0228314    .2270932    .3170061
-----------------------------+------------------------------------------------
              var(Residual)  |   .7676996   .0094288    .7494401    .7864038
------------------------------------------------------------------------------
LR test vs. linear regression: chibar2(01) =  2675.02 Prob >= chibar2 = 0.0000
```

estat icc

```
Intraclass correlation

------------------------------------------------------------------------------
              Level |      ICC   Std. Err.    [95% Conf. Interval]
--------------------+---------------------------------------------------------
           schoolid |   .2589837   .0165613    .2278667    .2927389
------------------------------------------------------------------------------
```

Is this model 'better'?

From the output we can see that:

- There are 13,646 observations (students) that belong to 356 groups (schools).
- There is a minimum of 3 students per school and a maximum of 57 students per school, with the mean around 38.
- The log likelihood is –18,024.825.
- _cons is –0.037, which is the mean standardized reading score for the sample. This is different from the model that used region as the Level 2 variable because now we are examining the average effect across the different schools, rather than regions.

- var(_cons) is 0.268 and is the between-group variance, or the random intercept of the Level 2 variable.
- var(Residual) is 0.768 and is the between-student within-group (Level 1) variance.
- rho, the ICC, is 0.259. Therefore, 25.9% of the variation in standardized reading score can be attributed to differences between schools.
- The likelihood ratio test of sigma_u = 0 has a value of 2,675.02 with a corresponding *p* value stated as 0.000, which gives us strong evidence in favour of rejecting the null. In other words, there is evidence that using a random intercept model helps explain the variance in standardized reading scores, even in the absence of any independent variables.

Comparing ICCs

The ICC reported for this model is 0.259. This value is over four times as high as the ICC associated with region, which was 0.058. It would therefore seem that variance between schools is a much more important explanatory feature of standardized reading scores than variance between regions.

If we create a caterpillar plot for this output (using the same technique as before, but substituting *schoolid* for *region*), we get a much more filled-out plot that actually resembles its insect namesake, due to the greater number of schools (Figure 2.4). There is one distinct outlier at the bottom left of the graph that comes from a very small school.

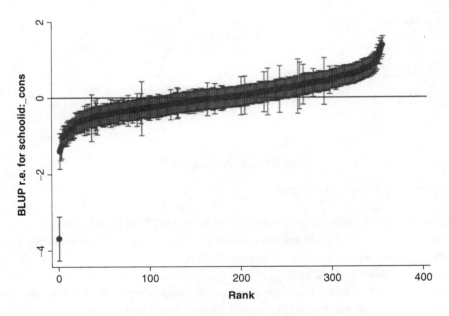

Figure 2.4 Standard error bar chart of relative random intercept values for school null model

As with the previous model that used region as the grouping variable, we can also tabulate the random intercept and its standard error for each school. In Table 2.10 we show only the first ten schools in the data set.

Table 2.10 Stata output of ranked random intercept values by school (first ten by rank)

```
+----------------------------------------+
| schoolid       u0         u0se    rank |
+----------------------------------------+
| 132      -3.685839    .2943393       1 |
| 154      -1.397695    .2353393       2 |
| 264      -1.336552    .1528458       3 |
|  51      -1.258072    .1690568       4 |
| 177      -1.045061    .1172304       5 |
|----------------------------------------+
| 165      -1.008524    .1322986       6 |
| 186      -.9951066    .1405521       7 |
| 249      -.9944479    .1370692       8 |
| 230      -.9723195    .1293817       9 |
| 149      -.949994     .1552259      10 |
+----------------------------------------+
```

We now have sufficient evidence that we have enough variation in our models to warrant multilevel modeling. In the next steps, we demonstrate how to build the models by adding independent variables.

ADDING LEVEL 1 EXPLANATORY VARIABLES

To add an independent variable to the model, the equation changes from the one for the null model to:

$$y_{ij} = \beta_0 + \beta_1 X_{1ij} + \mu_j + e_{ij}$$

The new part of the equation is $\beta_1 X_{1ij}$, which represents the independent variable. β_1 is the coefficient to be estimated and X_{1ij} is the observed value of the independent variable. If we had additional independent variables, they would be denoted by $\beta_2 X_{2ij}$, $\beta_3 X_{3ij}$, etc.

Figure 2.5 shows the random intercept model with one independent variable in a similar fashion to that of the null model in Figure 2.2 in that:

- β_0 is the overall mean of the dependent variable across all groups;
- μ_1 is the random intercept for Group 1 and μ_2 is the random intercept for Group 2;
- e_{ij} is the residual, which is the difference between the ith student's standardized reading score and their group's regression line.

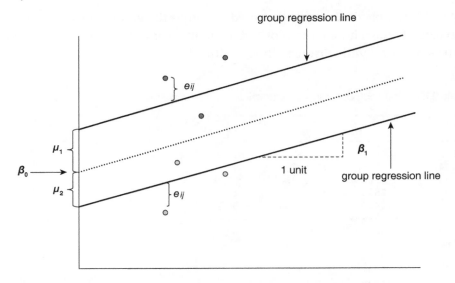

Figure 2.5 Graphical representation of a random intercept model with two groups

As the coefficient for the independent variable is 'fixed' to be the same for all groups the regression lines for all groups are parallel.

So far we have only looked at the null model with no independent variables in order to ascertain whether we should proceed with multilevel modeling. After comparing the region and school groupings, it has been shown that school explains considerably more variance in reading scores than region. Therefore, we will continue the worked example with *schoolid* as the Level 2 grouping variable. We are staying with two-level models for this example, but it would also make sense to examine a three-level model with these data which nested students within schools and then schools nested within regions.

As with ordinary regression, in order to have a theoretically sound model, the covariates that are associated with explaining reading scores need to be added to the model. We can add variables measured at Level 1 or Level 2, but first we will add variables measured at the student level (Level 1). Parental occupational status (*pos*) has been demonstrated to be a particularly strong predictor of student achievement, so we will add it to the model.

As it is easier to interpret the intercept when variables have a meaningful zero, in the forthcoming models we will use the centred *pos* variable created earlier in this chapter so that the mean is equal to zero. The estimations will be done only using **mixed**, but they can also be done using **xtreg**. The independent variable of interest, *cen_pos*, is included on the left-hand side of the command, as we are interested in its fixed effect:

```
mixed z_read cen_pos || schoolid:
```

The output is in Table 2.11.

Table 2.11 Stata output for school groups multilevel model of standardized reading scores regressed on parental occupational status using `mixed` command

```
Mixed-effects ML regression                Number of obs      =       13646
Group variable: schoolid                   Number of groups   =         356

                                           Obs per group: min =           3
                                                          avg =        38.3
                                                          max =          57

                                           Wald chi2(1)       =      758.81
Log likelihood = -17659.05                 Prob > chi2        =      0.0000

------------------------------------------------------------------------------
     z_read |      Coef.   Std. Err.      z    P>|z|     [95% Conf. Interval]
------------+-----------------------------------------------------------------
    cen_pos |   .0135055   .0004903    27.55   0.000     .0125446    .0144665
      _cons |  -.0290063   .0252515    -1.15   0.251    -.0784983    .0204857
------------------------------------------------------------------------------


------------------------------------------------------------------------------
  Random-effects Parameters  |   Estimate   Std. Err.    [95% Conf. Interval]
-----------------------------+------------------------------------------------
schoolid: Identity           |
                 var(_cons)  |   .2046757   .0181507     .172021    .2435294
-----------------------------+------------------------------------------------
               var(Residual) |   .7314819   .0089899     .7140726    .7493156
------------------------------------------------------------------------------
LR test vs. linear regression: chibar2(01) =  1891.60 Prob >= chibar2 = 0.0000
```

estat icc

Residual intraclass correlation

```
------------------------------------------------------------------------------
                       Level |        ICC Std. Err.     [95% Conf. Interval]
-----------------------------+------------------------------------------------
                    schoolid |   .2186338  .0153693     .1900051    .2502435
------------------------------------------------------------------------------
```

The coefficient for the centred parental occupational status variable (*cen_pos*) is interpreted in the same way that it would be in an ordinary regression. We can see that the value is statistically significant given its p value and also because the coefficient is several times larger than its standard error. Therefore, a one-unit increase in parental occupational status is associated with a 0.014 increase in standardized reading score.

In this first output with an independent variable it is worth discussing some of the differences between the output for the 'fixed' part and the random effects parameters. The output for the 'fixed' part looks very similar to that from most regression estimations with coefficients, standard errors, z values, p values and confidence intervals. These are interpreted and used in the same way as with other regression results, with the coefficient giving an estimate of the size and direction of the effect, and then the standard errors, z values, p values and confidence intervals giving an indication of the statistical significance and the precision of the coefficient. This part of the output should be familiar to you if you have used Stata to run single-level OLS regressions.

In the random effects parameters section of the output the random intercept is presented as a variance (or standard deviation, depending on your preferences) along with a standard error and a confidence interval. You might be wondering why we would be interested in the standard error and confidence interval of a variance or standard deviation, which are already distributional statistics. At this stage of your introduction to these models, we would advise you not to get too bogged down in the technicalities. However, it is worth noting two things: first, that the reported standard error should not be used in the same way to calculate a t or z value as you might for an independent variable coefficient; and second, that they do give an indication of the precision of the estimated variance or standard deviation. However, these confidence intervals do have a use as they provide some evidence that there is sufficient variation in the random parameter – in this example, the intercept – and so can be modeled in this way. We suggest that this evidence not be taken on its own but supplemented with a model **lrtest** (see after Table 2.18). In the random intercept model there are only two parameters in this section of the output, but in the next chapter this section of the output expands quite dramatically.

Replacing the constant (β_0) and independent variable coefficient (β_1) with the values from the Stata output gives:

$$y_{ij} = -0.029 + 0.014X_{1ij} + \mu_j + e_{ij}$$

The coefficient for parental occupational status (β_1) is 0.014 and is restricted to be the same for all schools, so it represents the effect of parental occupational status on standardized reading scores in the whole sample. However, the intercepts are allowed to vary and can be different for every school. Let's examine how these regression lines look.

```
predict predscore, fitted

egen justone=tag(schoolid cen_pos)

sort schoolid cen_pos

twoway connected predscore cen_pos if justone==1&schooled<11, ///

    connect(ascending) msize(vtiny)
```

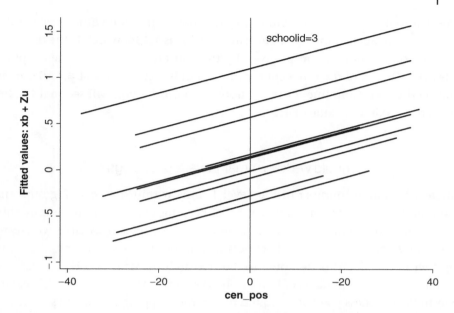

Figure 2.6 School regression lines for standardized reading scores on parental occupational status (first ten by *schoolid*)

The output in Figure 2.6 shows a series of lines that are parallel, but have different intercepts. This shows that this model constrains the coefficient for parental occupational status to be the same across all schools but allows the intercept for each school to vary. If we follow the same process as with the null model we can calculate the random intercept for each school in this new model with one independent variable. In Table 2.12 we show the first ten schools only, which are the same schools as those in Figure 2.6.

Table 2.12 Stata output of ranked random intercept values by school

```
+-----------------------------------------+
| schoolid       u0       u0se    rank   |
+-----------------------------------------+
|    3       1.134393   .1179551    356   |
|    4        .7455747  .1190932    348   |
|    8        .6022224  .1202649    338   |
|    6        .1894184  .1214718    244   |
|    5        .1726535  .1202649    239   |
+-----------------------------------------+
|   10        .162171   .125323     237   |
|    7        .0177733  .1239988    188   |
|    1       -.0586444  .1295653    154   |
|    2       -.2488444  .1342699     89   |
|    9       -.3347006  .125323      56   |
+-----------------------------------------+
```

We will look at one of the schools in more detail: the one with *schoolid* = 3 in Table 2.12 and labelled in Figure 2.6. The *u0* value is 1.134, which is the deviation from the total sample intercept given by the constant (–0.029) in the output in Table 2.11. Therefore the intercept for the school is –0.029 + 1.134 = 1.105. If you examine the regression line for this school in Figure 2.6 you will see that the line crosses the *y*-axis at a value of 1.105.

Should we keep small Level 2 groupings?

Is there a minimum number of the Level 1 units within the Level 2 groups that we should be aware of? You may be wary of keeping schools that have only three observations, for example. An examination of the *schoolnum* variable we created earlier indeed shows there are 11 schools with under ten observations. Should we drop them? You might be tempted to, but research by Maas and Hox (2005) has indicated that small numbers of Level 1 observations are not detrimental to parameter estimates. Unless you have a sound reason for dropping them, you should not.

Let us compare the results:

	Null model (Table 2.9)	+ cen_pos (Table 2.11)	% change
School-level variance	0.268	0.205	–23%
Student-level variance	0.768	0.731	–5%
ICC (rho)	0.259	0.218	–19%

Adding the independent variable *cen_pos* to the model resulted in the school-level variance changing from 0.268 to 0.205, representing a 23% decrease (1 – (0.205/0.268)). The student-level variance went from 0.768 to 0.731, a decrease of just 5% (1 – (0.731/0.768)). This suggests that the distribution of parental occupational status varies considerably among schools. Adding a Level 1 independent variable will usually reduce Level 1 variance as well as the total variance of the model. In terms of Level 2 variance, it may increase, decrease, or stay the same, depending upon how the variable of interest behaves across Level 2 groupings.

The ICC for the model with the independent variable *cen_pos* was 0.218, while the ICC for the null model was 0.259. By adding *cen_pos* to the model, we reduced the amount of explained variance that was due to differences between schools.

Adding a dichotomous independent variable

One important determinant of reading scores is gender. In general, females tend to have higher reading scores across the OECD, and Australia is no exception.

Therefore, it is important to add gender into the estimation. The variable measuring gender is named *female*, where 1 denotes female and 0 denotes male. To add it to the estimation, we just include it on the left-hand side of the Stata command:

```
mixed z_read cen_pos female || schoolid:

estat icc
```

Table 2.13 Stata output for school groups multilevel model of standardized reading scores regressed on parental occupational status and gender using `mixed` command

```
Mixed-effects ML regression               Number of obs      =        13646
Group variable: schoolid                  Number of groups   =          356

                                          Obs per group: min =            3
                                                         avg =         38.3
                                                         max =           57

                                          Wald chi2(2)       =      1441.36
Log likelihood = -17341.834               Prob > chi2        =       0.0000

------------------------------------------------------------------------------
     z_read |      Coef.   Std. Err.      z    P>|z|     [95% Conf. Interval]
------------+-----------------------------------------------------------------
    cen_pos |   .0134602   .0004789    28.10   0.000     .0125215    .014399
     female |   .4082597   .0160169    25.49   0.000     .3768671   .4396522
      _cons |   -.231443   .0264178    -8.76   0.000    -.2832208  -.1796651
------------------------------------------------------------------------------


------------------------------------------------------------------------------
  Random-effects Parameters  |   Estimate   Std. Err.    [95% Conf. Interval]
-----------------------------+------------------------------------------------
schoolid: Identity           |
                  var(_cons) |   .2047217   .0181184     .1721197    .243499
-----------------------------+------------------------------------------------
               var(Residual) |    .697465   .0085724     .6808641   .7144706
------------------------------------------------------------------------------
LR test vs. linear regression: chibar2(01) =  1960.21 Prob >= chibar2 = 0.0000

. estat icc

Residual intraclass correlation

------------------------------------------------------------------------------
                       Level |       ICC  Std. Err.    [95% Conf. Interval]
-----------------------------+------------------------------------------------
                    schoolid |   .2269173   .0157518     .1975322   .2592613
------------------------------------------------------------------------------
```

The results for gender are large and statistically significant (Table 2.13). Being female increases standardized reading scores by 0.408. We can compare the Level 2 and Level 1 variances again:

	Null model (Table 2.9)	+ cen_pos (Table 2.11)	+ female (Table 2.13)	% change (from Table 2.11 to Table 2.13)
School-level variance	0.268	0.205	0.205	0
Student-level variance	0.768	0.731	0.697	–5%
ICC (rho)	0.259	0.218	0.227	+4%

In this case, we can see that adding gender to the model has no effect on the school-level variances and decreases Level 1 variance by only 5%. The ICC also does not change much, indicating that the explained variance at the school level is not affected by the inclusion of this factor.

ADDING LEVEL 2 EXPLANATORY VARIABLES

In two-level multilevel models, we can include variables at both levels. In our example we have variables that measure individual characteristics of students, but we can add variables that measure the characteristics of their schools. One possible independent variable of interest is school size, which is called *schsize* in the data:

```
egen justone=tag(schoolid)
su schsize if justone==1
```

```
    Variable |  Obs        Mean      Std. Dev.      Min     Max
-----------------+--------------------------------------------------
     schsize |  356      863.7191    410.2669       10     2835
```

Summarizing the variable tells us that the 356 schools range in size from 10 to 2,835 students, with a mean school size of around 864 students. In keeping with the previous practice of wanting meaningful zeros for our variables, we will centre this variable as well because zero is not a plausible value for this variable – how could a school have zero students? We use the sample mean here but you may have good reasons to use another figure, such as making your analysis comparable with other studies.

```
gen cen_size = schsize - r(mean)
```

As we have just used the **summarize** command for the variable *schsize* (the **if justone==1** part is to use the appropriate denominator for the number of schools, not the number of students) then we can use the scalar for the mean and check the new variable:

```
su cen_size if justone==1
```

```
    Variable |    Obs           Mean    Std. Dev.          Min         Max
-------------+----------------------------------------------------------------
    cen_size |    356       -7.76e-06    410.2669     -853.7191    1971.281
```

The new variable *cen_size* can be added to the command in the same way that Level 1 variables were added:

```
mixed z_read cen_pos female cen_size || schoolid:

estat icc
```

Table 2.14 Stata output for school groups multilevel model of standardized reading scores regressed on parental occupational status, gender and school size using `mixed` command

```
Mixed-effects ML regression                 Number of obs       =        13646
Group variable: schoolid                    Number of groups    =          356

                                            Obs per group: min  =            3
                                                           avg  =         38.3
                                                           max  =           57

                                            Wald chi2(3)        =      1494.11
Log likelihood = -17322.401                 Prob > chi2         =       0.0000

------------------------------------------------------------------------------
      z_read |      Coef.   Std. Err.      z    P>|z|     [95% Conf. Interval]
-------------+----------------------------------------------------------------
     cen_pos |   .0133781   .0004791    27.92   0.000      .012439    .0143172
      female |   .4095749   .0159999    25.60   0.000     .3782156    .4409341
    cen_size |   .0003781    .000059     6.40   0.000     .0002624    .0004938
       _cons |  -.2360258   .0250803    -9.41   0.000    -.2851823   -.1868692
------------------------------------------------------------------------------

------------------------------------------------------------------------------
  Random-effects Parameters  |   Estimate   Std. Err.    [95% Conf. Interval]
-----------------------------+------------------------------------------------
schoolid: Identity           |
                 var(_cons)  |   .1801154   .0161573     .1510753    .2147377
-----------------------------+------------------------------------------------
               var(Residual) |   .6975789   .0085741     .6809747    .7145879
------------------------------------------------------------------------------
LR test vs. linear regression: chibar2(01) = 1765.34 Prob >= chibar2 = 0.0000

. estat icc

Residual intraclass correlation

------------------------------------------------------------------------------
                       Level  |        ICC   Std. Err.    [95% Conf. Interval]
-----------------------------+------------------------------------------------
                    schoolid  |   .2052143   .0148437     .1776439    .2358365
------------------------------------------------------------------------------
```

The results in Table 2.14 show a coefficient for *cen_size* of 0.0004, which is small but significant. Remember that this variable measures numbers of students per school, and this coefficient would represent the addition of one student. So, for every one-student increase in school size standardized reading scores increase by 0.0004. The intercept (–0.236) is interpreted as the standardized reading score for a male (female = 0) student with the mean parental occupational status and at a school with the mean number of students.

We can now see that by adding school size the school-level variance decreased by 14% and the ICC decreased by 11%, but the Level 1 variance did not change:

	Null model (Table 2.9)	*+ cen_pos + female* (Table 2.13)	*+ cen_size* (Table 2.14)	*% change* (from Table 2.13 to Table 2.14)
School-level variance	0.268	0.205	0.180	–14%
Student-level variance	0.768	0.697	0.697	0
ICC (rho)	0.259	0.227	0.205	–11%

Adding a Level 2 variable that is then significant in the model would be expected to decrease the variance left to explain by school differences.

GROUP MEAN CENTRING

Up to this point, we have centred variables about the mean of the total sample, which is often referred to as *grand* mean centring. Sometimes it is appropriate to centre Level 1 variables around their Level 2 group mean rather than the overall mean, which is called *group* mean centring. For example, for parental occupational status a grand mean centred variable would measure the difference from the mean parental occupational status of all students (which we have done in the examples so far), while a group mean centred variable would measure the difference from the mean parental occupational status of students within a group (in our current example, the school). Group mean centring of parental occupational status (*pos*) by school can be accomplished as follows:

```
egen posmeans = mean(pos), by(schoolid)
gen grcen_pos = pos - posmeans
```

Adding this to the estimation:

```
mixed z_read grcen_pos female cen_size || schoolid:
```

we obtain the output in Table 2.15.

Table 2.15 Stata output for school groups multilevel model of standardized reading scores regressed on parental occupational status (group mean centred), gender and school size using `mixed` command

```
Mixed-effects ML regression                    Number of obs      =      13646
Group variable: schoolid                       Number of groups   =        356

                                               Obs per group: min =          3
                                                            avg =       38.3
                                                            max =         57

                                               Wald chi2(3)       =    1399.76
Log likelihood = -17358.681                    Prob > chi2        =     0.0000
```

z_read	Coef.	Std. Err.	z	P>|z|	[95% Conf. Interval]	
grcen_pos	.0127704	.0004834	26.42	0.000	.011823	.0137178
female	.4100387	.0160322	25.58	0.000	.3786161	.4414612
cen_size	.0004684	.0000662	7.08	0.000	.0003387	.0005982
_cons	-.2440981	.0279292	-8.74	0.000	-.2988384	-.1893579

Random-effects Parameters	Estimate	Std. Err.	[95% Conf. Interval]	
schoolid: Identity				
var(_cons)	.2336251	.0200191	.1975063	.2763492
var(Residual)	.6969869	.0085612	.6804076	.7139702

```
LR test vs. linear regression: chibar2(01) = 2568.15 Prob >= chibar2 = 0.0000
```

The intercept is now –0.244, which corresponds to the predicted standardized reading score for students who are male (female = 0) in a school with the mean number of students, and with parental occupation at the mean for the school that the student attends (group mean centred). The coefficient for the group mean centred variable measuring parental occupational status, *grcen_pos*, is 0.013 and can be interpreted as saying that a one-unit increase in parental occupational status is associated with a 0.013-unit increase in standardized reading scores. This group centring allows for estimates of student-level effects *within* each group.

Some important points need to be made about grand mean and group mean centring. Both have their advantages and disadvantages, but you should be guided in your choice between them primarily by your research questions. Enders and Tofighi (2007) provide a fuller discussion of this issue and some rules of thumb for deciding which method of centring to choose. Finally, just because you *can* do it does not mean you *should* add increasing degrees of complexity to your model without sound theoretical reasons. Your model should correspond as closely as possible to your theory.

INTERACTIONS

Just as in ordinary regression, interactions can be added to multilevel models. The major difference is that with multilevel models interactions between variables at different levels can be included. You may interact Level 1 variables with each other, and likewise Level 2 variables. But you can also create interactions between variables at different levels if you believe that individual student characteristics operate differentially according to school characteristics (i.e. Level 1 * Level 2).

Interactions are added to a model by including the product of the variables of interest alongside the main (also call 'direct') effects of the variables. If we want to examine if there is an interaction effect, we must include the two individual variables as well as their product (*variable1 * variable2*) in the model. The interpretation of the interaction cannot occur without the concurrent interpretation of the main effects (Hox, 2010). Regardless of whether you are looking at interactions in a single-level model or a multilevel model, or whether your interactions are across the same or different types of levels, it is important to remember to include both the main and interaction effects in your model, even if the main effects fail to reach statistical significance (Hox, 2010).

In addition to retaining both the main and interaction effects, it is important to recognize that the interpretation of the main effect changes once an interaction term is added to the model. When there is an interaction in our model between two variables, $X1$ and $X2$, the main effect of $X1$ is the value of the coefficient when $X2$ is equal to zero. And, conversely, the main effect of $X2$ is the value of the coefficient when $X1$ is equal to zero. This is yet another reason why it is important that the zeros in our variables have a meaningful interpretation. If zero is beyond the range of plausible values, then the coefficient has no meaningful interpretation. Recall from earlier in this chapter that we can make our zeros meaningful through the process of centring. If variables are rescaled so that they are mean centred and then added to a model with an interaction ($X1 * X2$), the main effect of $X1$ is now interpreted as the value of the coefficient when $X2$ is *at its mean value*.

The interpretation of interaction effects varies in difficulty depending on the level of measurement of the composite variables. It is easiest when one or both of your variables are dichotomous and most complex when both of the variables are continuous. In this chapter, only a simple overview is provided. For detailed discussions about interactions in multilevel models, see Hox (2010) and, more generally, Brambor et al. (2006).

Level 1 dichotomous by Level 2 continuous

The first example we will consider is the interaction of gender (Level 1, dichotomous) with the percentage of girls in a school (Level 2, continuous). Gender has been dummy coded into *female*, where female = 1 and male = 0. The percentage of females in a school (*pcgirls*) is represented by the proportion of girls in a school, which ranges from 0 (no girls) to 1 (all girls). We would pursue such an interaction if we had a theoretical basis for believing that girls' reading scores are influenced by the gender composition of their school – it may be reasonable to predict that girls do better in schools where there are more girls.

```
su pcgirls if justone==1
```

```
    Variable |        Obs        Mean    Std. Dev.        Min         Max
-------------+----------------------------------------------------------
     pcgirls |        356     .495486     .2055885          0           1
```

Descriptive statistics show that the mean percentage of girls in a school is around 49.5%. This variable has a meaningful zero, which indicates an all-boys school. However, because zero is uncommon, representing the 25 all-boys schools, it may make more sense to grand mean centre the variable so that it represents the mean percentage of girls in a school.

```
gen cen_girls = pcgirls - r(mean)
```

We also need to create an interaction term to add to the model. In this first example of interaction effects we run through the process 'manually', creating interaction terms and graphing. In the next example we show you some of the other ways to do this in Stata.

```
gen femaleXcen_girls=female*cen_girls
```

We now add the main effects (*female* and *cen_girls*) and the interaction (*femaleXcen_girls*) to the left-hand side of the **mixed** command. Note that to keep interpretation simple, we only include these three variables in the model.

```
mixed z_read female cen_girls femaleXcen_girls || schoolid:
```

The results shown in Table 2.16 reveal that both of the main effects and the interaction term are significant. The coefficient for *female* is 0.421 when *cen_girls* is at its mean (0). The coefficient for *cen_girls* is –0.890 when *female* is equal to zero (male).

Table 2.16 Stata output for school groups multilevel model of standardized reading scores regressed on gender and percentage of girls in school with interaction term

```
Mixed-effects ML regression              Number of obs      =        13646
Group variable: schoolid                 Number of groups   =          356

                                         Obs per group: min =            3
                                                        avg =         38.3
                                                        max =           57

                                         Wald chi2(3)       =       650.00
Log likelihood = -17708.062              Prob > chi2        =       0.0000


-----------------------------------------------------------------------------
           z_read |   Coef.   Std. Err.    z    P>|z|   [95% Conf. Interval]
------------------+----------------------------------------------------------
           female |  .4209452  .0166701  25.25  0.000   .3882724    .453618
        cen_girls | -.8904694  .1852812  -4.81  0.000  -1.253614  -.5273249
femaleXcen_girls |  1.328636  .2427706   5.47  0.000   .852814    1.804457
            _cons | -.3018326  .0301456 -10.01  0.000  -.3609169  -.2427483
-----------------------------------------------------------------------------


-----------------------------------------------------------------------------
  Random-effects Parameters |  Estimate  Std. Err.   [95% Conf. Interval]
----------------------------+------------------------------------------------
schoolid: Identity          |
                var(_cons)  |  .2315181  .0204087   .1947826    .2751817
----------------------------+------------------------------------------------
              var(Residual) |  .734687   .0090305   .717199     .7526014
-----------------------------------------------------------------------------
LR test vs. linear regression: chibar2(01) = 2272.32 Prob >= chibar2 = 0.0000
```

The fixed part of the estimation is:

$$Z\_read = -0.302 + 0.421\ female_{ij} - 0.890 cen\_girls_{ij} + 1.329\ femaleXcen\_girls_{ij}$$

To help understand the interaction effect, we can substitute in values for different instances and then graph it.

For males in schools with the mean proportion of girls (zero on the centred variable), this value is the same as the constant because there is a zero on the *female* coefficient and a zero on the *female* component of the interaction:

$$Z\_read = -0.302 + 0.421(0) - 0.890(0) + 1.329(0 \times 0) = -0.302$$

For females in schools with the mean proportion of girls (i.e. they get a one on the *female* coefficient and a one on the *female* component of the interaction), this value is:

$$Z\_read = -0.302 + 0.421(1) - 0.890(0) + 1.329(1 \times 0) = 0.119$$

For females in schools with 10% more than the average number of girls (i.e. 0.10 replaces the *cen_girls* value and the corresponding part of the interaction), this value is:

$$Z\_read = -0.302 + 0.421(1) - 0.890(0.1) + 1.329(1 \times 0.1) = 0.163$$

Thus, there is evidence that girls do better, the more girls there are in their school. This can be more clearly illustrated by creating a graph of this interaction (Figure 2.7):

```
predict girls, fitted
twoway lfit girls cen_girls if female == 1 || ///
       lfit girls cen_girls if female ==0, ///
       legend(order(1 "girls" 2 "boys" )) ///
       ytitle(predicted reading) ///
       lpattern(solid) lpattern(dash)
```

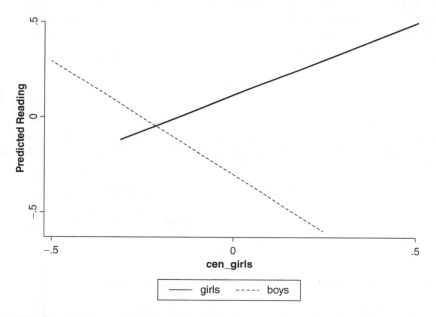

Figure 2.7   Graphical representation of interaction between gender and percentage of girls in a school on standardized reading scores

Level 1 continuous by Level 2 continuous

Conceptually, there are many ways to construct interaction terms in multilevel models. In sociology and education research, there is sometimes an interest in examining the impact an independent variable has on a dependent variable at both Level 1 and Level 2. For example, you might have a hypothesis that the socioeconomic status of students impacts on academic achievement differentially, depending on the average socioeconomic status of students attending the school. Perhaps the socioeconomic status of students is more important in schools where the average student has high socioeconomic status. These types of models are sometimes called contextual or compositional models (Enders and Tofighi, 2007).

In this example we examine the effects of parental occupational status and teacher–student ratio (TSR, variable *stratio*) on standardized reading scores for boys only. In our fuller analysis of the data we discovered a three-way interaction between parental occupational status, teacher–student ratio and gender. For this example we stay with a two-way interaction by using the subsample of boys only. We have centred the TSR variable (*cen_ratio*), but as we are dealing with a Level 2 variable we must, again, use the number of schools in the denominator to do this rather than the number of students.

Now we create a model that includes the main effects and the interaction term (note that other independent variables are not included to simplify the interpretation of the output). In this command we show the use of factor variables in Stata and how to interact them without needing to create additional variables manually. Putting **##** between the two variables tells Stata to estimate both main effects and the interaction. The **c.** before the variables indicates to treat these variables as continuous.

```
mixed z_read c.cen_pos#c.cen_ratio if || schoolid: ///
  if female==0
```

Table 2.17 Stata output for school groups multilevel model of standardized reading scores regressed on parental occupational status and school teacher–student ratio with interaction term, boys only

```
Mixed-effects ML regression            Number of obs     =        6879
Group variable: schoolid               Number of groups  =         323

                                       Obs per group: min =           2
                                                      avg =        21.3
                                                      max =          51

                                       Wald chi2(3)      =      395.44
Log likelihood = -9019.2533            Prob > chi2       =      0.0000
```

z_read	Coef.	Std. Err.	z	P>|z|	[95% Conf. Interval]	
cen_pos	.01362	.00069	19.68	0.000	.01226	.01498
cen_ratio	-.00482	.01035	-0.47	0.641	-.02513	.01547
c.cen_pos#c.cen_ratio	-.00075	.00030	-2.48	0.013	-.00136	-.0001
_cons	-.24635	.02637	-9.34	0.000	-.29804	-.19467

Random-effects Parameters	Estimate	Std. Err.	[95% Conf. Interval]	
schoolid: Identity				
var(_cons)	.18255	.01769	.1509	.22075
var(Residual)	.74205	.01296	.71708	.76790

```
LR test vs. linear regression: chibar2(01) = 924.06 Prob >= chibar2 = 0.0000
```

The output in Table 2.17 reveals that the main effect for parental occupational status is significant and the interaction, while small, is also significant. The main effect for TSR is non-significant.

Unlike with nominal by continuous interactions, understanding what is going on in continuous by continuous interactions takes a little bit more work. A typical strategy is to break the Level 2 variable into categories representing high, medium, and low values of that variable. Here, we will break the centred school TSR variable (*cen_ratio*) into tertiles, or three equally sized parts of the distribution, and create a new variable *tertiles*. As we are dealing with a Level 2 variable again we must use the number of schools to do this rather than the number of students.

We can now graph the individual parental occupational status variable with the predicted reading score, using these tertiles as group markers to represent the predicted gradients for groupings of schools.

```
quietly margins, at(cen_pos=(-20(5)20)) over(tertiles)
marginsplot, noci legend(order(1 "Low TSR school" ///
    2 "Med TSR school" 3 "High TSR school"))
```

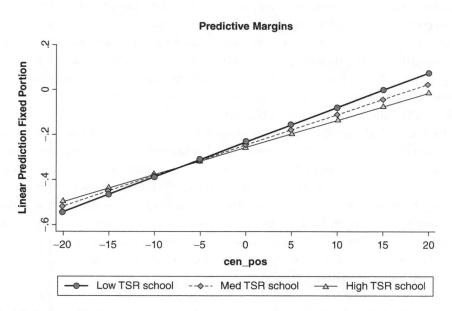

Figure 2.8 Graphical representation of interaction between parental occupational status and teacher–student ratio on standardized reading scores, boys only

Unsurprisingly as the interaction coefficient was small, Figure 2.8 shows only slight differences between the three lines, one each for low, medium and high TSR schools. The graph shows that students with higher parental occupational status achieve higher standardized reading scores in schools with lower TSRs. For those boys at, and below, the mean parental occupational status the differences are smaller. Another way of interpreting the interaction is that the effect of parental occupational status is greater (steeper slope) in schools with low TSRs.

MODEL FIT

Most model fit statistics for random intercept models are not provided by the default settings of **mixed**, but can be obtained quite easily by requesting **estat ic** after your estimation. The default settings produce the log likelihood (LL) in the output. The Akaike information criterion (AIC) and the Bayesian information criterion (BIC) are produced by the **estat ic** command.

The model fit statistics are generally measuring whether the 'lack of fit' of a model is improving. Remember that the difference between an observed score (in the data) and the predicted score (from the estimation) produces a residual. The sum of all the squared residuals (or errors) is the sum of the lack of a model's fit. A regression line is the line that corresponds to the 'least squared errors'. The smaller these errors are, the better the model. So if we continue to improve the fit of the model, the sum of the overall errors should decrease. The log likelihood (LL) can be thought of as the lack of fit of a model and is reported in the standard **mixed** output. The log likelihood is often converted to a fit statistic called the reduction in deviance that is equal to –2(log likelihood of Model 2 – log likelihood of Model 1) and usually follows a chi-square distribution.

The AIC and BIC are different indicators of fit. When evaluating a deviance, our objective is to decrease it. This, however, is not the only measure of a good model. AIC and BIC evaluate models in terms of their parsimony/complexity as well as their statistical fit. To grossly oversimplify, these fit statistics impose a penalty on having too many parameters (i.e. variables) in the model. The differences between AIC and BIC are beyond the scope of this book. In general, they are very similar, although the BIC 'penalizes' more for extra parameters than the AIC. To evaluate model fit, we want the AIC and BIC scores to decrease – that is to say, AIC and BIC scores are lower for better-fitting models.

Let us run four of the models we have shown already in this chapter (Tables 2.9, 2.11, 2.13 and 2.14) and examine their fit statistics. Remember that Model 1 includes the independent variable *female*; Model 2 includes two independent variables *female* and *cen_pos*; and Model 3 adds the Level 2 variable *cen_size*.

We use the commands **estat ic** and **xtmrho** after the estimation of the model to get the estimates we need, including AIC and BIC. We repeat this three more times and use the **quietly** prefix to save producing the output in the results widow. We then use the **estimates store** command to save the model in Stata's memory, calling them null, model 1, model 2, and model 3 (you can call them whatever you want). We only show part of the output in Table 2.18. See the next section on *R*-squared for the explanation for why we use **xtmixed** in the commands below.

```
quietly xtmixed z_read || schoolid:

estat ic

xtmrho

estimates store null


quietly xtmixed z_read female || schoolid:

estat ic

xtmrho

estimates store model1


quietly xtmixed z_read female cen_pos || schoolid:

estat ic

xtmrho

estimates store model2


quietly xtmixed z_read female cen_pos cen_size || schoolid:

estat ic

xtmrho

estimates store model3
```

Table 2.18 Fit statistics from a series of models

Model	Null (Table 2.9)	Model 1 (Table 2.11)	Model 2 (Table 2.13)	Model 3 (Table 2.14)
AIC	36,055.7	35,452.5	34,693.7	34,656.8
BIC	36,078.2	35,482.6	34,731.3	34,701.9
LL	–18,024.8	–17,722.3	–17,341.8	–17,322.4

In terms of the AIC and the BIC values, you can see that they have decreased in each successive model, suggesting that our fit is improving. We can calculate the LL change for the null model to Model 1, Model 1 to Model 2, etc. by using the **lrtest** command:

```
. lrtest model1 null

Likelihood-ratio test                        LR chi2(1) = 605.13

(Assumption: null nested in model1)          Prob > chi2 = 0.0000


. lrtest model2 model1

Likelihood-ratio test                        LR chi2(1) = 760.85

(Assumption: model1 nested in model2)        Prob > chi2 = 0.0000


. lrtest model3 model2

Likelihood-ratio test                        LR chi2(1) = 38.87

(Assumption: model2 nested in model3)        Prob > chi2 = 0.0000
```

The chi-square values are on the right of the output along with the associated p value for the chi-square distribution. These tests also indicate that the models are getting 'better' as there is a significant change in the log likelihood each time.

WHAT ABOUT R-SQUARED?

You are likely familiar with using R^2 or adjusted R^2 for assessing the goodness of fit (technically, the proportion of variance explained in your dependent variable by your independent variables) of your single-level regression models. If random intercept models are just an extension of these models, then certainly R^2 must factor in here somehow, right? Well, you have joined one of the areas of debate in multilevel modeling. Some have come up with different ways of calculating an R^2 (Bryk and Raudenbush, 2002; Rabe-Hesketh and Skrondal, 2008; Snijders and Bosker, 1994, 1999), while others advocate not using these various R^2 and sticking to the other measures of fit (Kreft and de Leeuw, 1998). If you wish to include R^2 in your assessment of model fit then you have a couple of options for obtaining this information.

Rabe-Hesketh and Skrondal (2008) have suggested that the R^2 for a two-level model can be calculated, but that it must be broken down into its composite parts. Specifically, they suggest that:

$$\text{Overall } R^2 = \frac{\left[\begin{array}{l}\text{Level 2 variance (null) + Level 1 variance (null)} - [\text{Level 2 variance} \\ \text{(model of interest) + Level 2 variance (model of interest)}]\end{array}\right]}{\text{Level 2 variance (null)} + \text{Level 1 variance (null)}}$$

$$\text{Level 1 } R^2 = \frac{\text{Level 1 variance (null model)} - \text{Level 1 variance (model of interest)}}{\text{Level 1 variance (model of interest)}}$$

$$\text{Level 2 } R^2 = \frac{\text{Level 2 variance (null model)} - \text{Level 1 variance (model of interest)}}{\text{Level 2 variance (model of interest)}}$$

To calculate these Rabe-Hesketh and Skrondal (2008) R^2 values we need to use the values stored as scalars after the **xtmrho** command. If you envisage doing this many times, you might want to write your own small routine. At the time of writing, the **xtmrho** command only works properly after the **xtmixed** (rather than **mixed**) command. The **xtmixed** command runs perfectly well in Stata 13, so the commands for this section use **xtmixed**. Also just running the **xtmixed** command and storing estimates will not produce the right scalars; you need to run **xtmrho** after. Then these scalars are put into the simple computation of Level 1, Level 2 and overall R^2 values. We do this for the three models in Table 2.18,

```
estimates restore null

scalar a=e(var_u1)

scalar b=e(var_e)

estimates restore model1

scalar c=e(var_u1)

scalar d=e(var_e)

scalar x=[[scalar(a)+scalar(b)]-[scalar(d)+scalar(c)]]///
    [scalar(a)+scalar(b)]
```

and so on for Models 2 and 3. The final output is:

```
Model 1 Overall R2 is: .03228312

Model 2 Overall R2 is: .12917096

Model 3 Overall R2 is: .1528121
```

The calculations indicate that for Model 3 the proportion of variance explained at Level 1 is about 10%, while it is about 49% at Level 2. Thus, the model explains more at the school level than the individual level. Overall, the model explains about 15% of the total variance (Table 2.19). It should be obvious at this point that the overall R^2 is *not achieved* by simply adding the Level 1 R^2 and the Level 2 R^2 together!

Table 2.19 Fit statistics from previous models, including R^2

Model	Null (Table 2.9)	Model 1 (Table 2.11)	Model 2 (Table 2.13)	Model 3 (Table 2.14)
AIC	36,055.7	35,452.5	34,693.7	34,656.8
BIC	36,078.2	35,482.6	34,731.3	34,701.9
LL	–18,024.8	–17,722.3	–17,341.8	–17,322.4
$R^{2\S}$	–	0.032	0.129	0.153

§ R^2 from Rabe-Hesketh and Skrondal (2008)

Table 2.20 Various calculations of R^2 by level

R^2	Bryk and Raudenbush (2002)	Snijders and Bosker (1994)	Rabe-Hesketh and Skrondal (2008)
Level 1	0.091	0.153	0.101
Level 2	0.329	0.308	0.489
Overall	–	–	0.153

Another option for generating R^2 for these models is to install the mlt package – **capture ssc install mlt** – and then use the command **mltrsq**. This is another command that, at the time of writing, will only work after the **xtmixed** command, and it produces R^2 values from Bryk and Raudenbush (2002) and

Snijders and Bosker (1994). In Table 2.20 we show the different R^2 produced from both of these approaches for Model 3 from Table 2.20.

DIAGNOSTICS

You will likely remember from your introductory statistics classes the emphasis placed on regression diagnostics. To refresh your memory, these are simply tests we perform in order to see if our model adheres to some of the assumptions of the models we are using. The 'assumptions' are a list of conditions under which our statistical tool operates properly. As mentioned earlier, one of the major reasons for using multilevel modeling is that nested data violate key assumptions of OLS regression.

Diagnostics in multilevel modeling mostly have to do with the analysis of residuals. In single-level regression the only residual is that representing the difference between the predicted and observed dependent variable ($\hat{y}$ or 'y-hat' and y, respectively). In multilevel modeling there are residuals at all levels of the analysis. Under the assumptions of multilevel modeling, all residuals have a mean of zero, are normally distributed, are homoscedastic, and are uncorrelated with any of the independent variables – sometimes called the exogeneity. These conditions will look familiar to you from your knowledge of single-level OLS regression. The whole gamut of statistical assumptions can be found in more detailed treatments of multilevel modeling (e.g. Snijders and Bosker, 1999). In this section we will focus on the ones that we can actually check.

The Level 1 residuals are analogous to those in single-level regression in that they are the difference between the predicted and observed values of the dependent variable. If we think about the models with only one independent variable, as they are easier to picture, the difference in these random intercept models is that each school has a separate regression line from which to calculate the residual rather than a single regression line for the whole sample. In random intercept models where the group regression lines are parallel to each other, the Level 2 residuals are the same as the random intercept values (μ_1 and μ_2 in Figure 2.5).

Both Level 1 and Level 2 residuals can be requested immediately after an estimation using the **predict** command. The Level 1 residuals are obtained from the **residuals** option and the Level 2 residuals are obtained from the **reffects** option. We run our last model first and then create histograms of both residuals.

```
mixed z_read female cen_pos cen_size || schoolid:

predict L1_resid, residuals

predict L2_resid, reffects

histogram L1_resid, normal title ///
   ("Histogram of Level 1 Residuals") saving(h1)

histogram L2_resid, normal title ///
   ("Histogram of Level 2 residuals") saving(h2)

qnorm L1_resid, title ///
   ("Qnorm Plot of Level 1 Residuals") saving(q1)

qnorm L2_resid, title ///
   ("Qnorm Plot Level 2 Residuals") saving(q2)

*combine all graphs into one output

graph combine h1.gph h2.gph q1.gph q2.gph
```

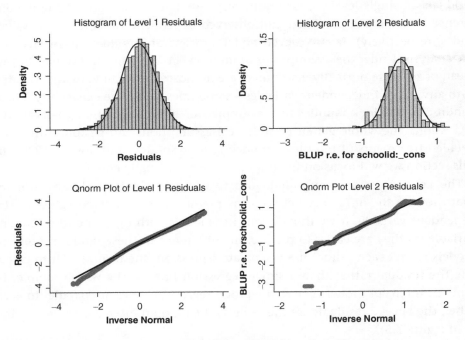

Figure 2.9   Histograms and quantile normal plots of Level 1 and Level 2 residuals

The histograms and quantile normal (qnorm) plots in Figure 2.9 give us very similar information. They are just two ways of looking at the distribution of our residuals. The histograms show that the Level 1 residuals are normally distributed and that the Level 2 residuals have notable outliers. In terms of the qnorm plot, we want it to fit the straight diagonal line as closely as possible. Again, we can see that there is indeed evidence of non-normality in the case of the Level 2 residuals. At this point, we may want to investigate these outliers. There is no straightforward treatment of outliers in multilevel data (or any type of data, for that matter). Interested readers are advised to consult detailed treatments of the matter, such as Langford and Lewis (1998).

A FURTHER ASSUMPTION AND A SHORT NOTE ON RANDOM AND FIXED EFFECTS

The assumption is that the correlation between the Level 2 (schools in our example) residuals and the independent variables is zero. This assumption is regardless of the models being random intercept or, in the following chapter, random coefficient models. Both random intercept and random coefficient models can be grouped as 'random effects' models, and the assumption is sometimes referred to as the 'random effects assumption'. See Clarke et al. (2015) for an accessible overview of the uses and limitations of multilevel (they call them hierarchical) fixed and random effects models. Clarke et al. (2015) compare the use of random effects and fixed effects models with nested data. As part of our introduction in Chapter 1 (Table 1.4) we specified a basic fixed effects model (using regions as the Level 2 groups), but these models cannot properly estimate the effects of Level 2 variables (regional characteristics in the example in Chapter 1). However, the Level 2 residuals can be correlated with the independent variables. We recommend that you read Clarke et al. (2015) – perhaps after working through Chapter 3 – and note that this difference in assumptions between fixed and random effects models also plays out when using panel data.

--- ## CHAPTER 2 TAKEAWAY POINTS ---

- Random intercept models are the most basic and most widely used types of multilevel models.
- In random intercept models, the intercept is allowed to vary for each Level 2 group.
- The general consensus is that you should have more than 20 Level 2 groups for a multilevel estimation.
- You should initially run a null model (with no covariates) to check how much variance in your dependent variance is explained by differences between your Level 2 groups.
- In the exploratory stage of your analysis, you should check the likelihood ratio test and intraclass correlation statistics to determine if multilevel modeling is necessary in your analysis. If there is no evidence of 'nesting effects' there is little reason to pursue multilevel models.

THREE

Random Coefficient Models: When intercepts and coefficients vary

CONTENTS

In the previous chapter, it was shown that single-level OLS regression models can be expanded to accommodate a theoretical assumption that there are different (or random) intercepts for each unit of a grouping, or cluster, or nesting factor. In the examples in Chapter 2, random intercepts were modeled for region and school groupings. Picturing a model for schools as the grouping factor with only one independent variable (easier to visualize in two dimensions), each school had its own intercept point on the y-axis, but the linear trajectory was identical for each unit. In other words, the regression lines for all the schools were parallel, with the same coefficient or slope, but they had different intercepts.

We can develop the multilevel model further so that these lines are allowed to have different slopes. We use the term *random coefficient models* but, as you can probably guess, there are other names for them. You should also be aware that random coefficient models also *assume random intercepts* in most cases. It is possible to have models with random coefficients and a fixed intercept, but these are few and far between. There are just not many research questions that suggest such relationships exist in the social world (i.e. random slopes emanating from a single fixed intercept). Therefore, you can be pretty certain that when people are talking about random coefficient models they are referring to models where both the intercepts and coefficients are allowed to vary across groups.

In Chapter 2, we described the equation for a null random intercept model:

$$y_{ij} = \beta_0 + \mu_j + e_{ij}$$

Remember that in this equation the 'random' part that allows the intercept to vary across groups (schools in most of our examples) is μ_j. We showed that this part of the equation is how much each school intercept varies from the overall sample intercept β_0.

Then we illustrated how random intercept models with independent variables, whose coefficients were constrained to be the same across all schools (the 'fixed' part, β_1), were another extension (see Figure 2.5 to review):

$$y_{ij} = \beta_0 + \beta_1 X_{1ij} + \mu_j + e_{ij}$$

The random coefficient model is one more extension, where we now have two random terms instead of just one:

$$y_{ij} = \beta_0 + \beta_1 X_{1ij} + \mu_{0i} + \mu_{1i} X_{ij} + e_{ij}$$

You may prefer to see the intercept and independent variable parts of the equation next to each other:

$$y_{ij} = \underbrace{\beta_0 + \mu_{0j}}_{} + \underbrace{\beta_1 X_{1ij} + \mu_{1j} X_{1ij}}_{} + e_{ij}$$

intercept   coefficient

Notice that in the random coefficient model, instead of μ_j, there is μ_{0j} representing the random intercept. The term $\mu_{1j} X_{1ij}$ represents the random coefficient. In a similar way to how μ_j captured the random intercept, the term $\mu_{1j} X_{1ij}$ captures the random coefficient by giving us the difference between each school's coefficient from the overall sample coefficient $\beta_1 X_{1ij}$ for each independent variable. We illustrate what these terms are modeling below.

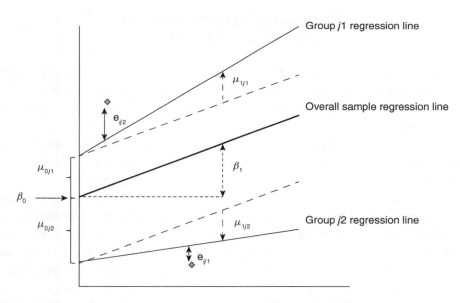

Figure 3.1 Illustration of the elements of a random coefficients model (adapted from Rabe-Hesketh and Skrondal, 2008)

In Figure 3.1 the heavy line in the middle represents the overall sample regression line with an overall sample coefficient (slope, if you prefer) β_1 and intercept β_0. The lighter lines above and below it represent the regression lines for two different groups ($j1$ and $j2$).

We have discussed random intercepts in the previous chapter, but to recap, there is the overall sample intercept (β_0) and the intercepts for the two groups with their differences from β_0, denoted by μ_{0j1} and μ_{0j2}, respectively.

The dashed grey lines are parallel to the overall sample regression line, and they are used to illustrate the difference between the overall sample coefficient (slope) and the group-specific coefficient (slope). The group represented by the top regression line has a larger coefficient (steeper slope) and so the difference from the overall sample (μ_{1j1}) is positive. The group represented by the lower regression line has a negative difference (μ_{1j2}) because its coefficient is smaller.

The residuals (illustrated by e_{ij1} and e_{ij2}) are the difference between the individual observation and the regression line for the group (group-specific predicted value) that the individual belongs to.

All these elements are part of the estimation of the random coefficients model.

GETTING STARTED WITH RANDOM COEFFICIENT MODELS

In a similar way to how we examined our data to see if a random intercept model was more appropriate than a single intercept one, we can first look at our data and determine if there is enough variation in the coefficients (slopes) to warrant further modeling in a multilevel framework.

There are a number of possible ways to go about this initial examination of the data, but here we start by running separate OLS regressions for each school and then examine results using some descriptive statistics and graphs.

We will continue with our example that investigates the effect of parental occupational status (the centred variable *cen_pos*) on standardized reading scores (*z_read*). The command **statsby** is useful for this procedure as it collects the value of the intercept and the coefficient for each regression and we tell Stata to save these results in a new data set, with one observation per school, which we will first inspect and then merge back into our working data for graphing. We called the saved data set ols.dta in this example.

```
statsby intercept=_b[_cons] coeff=_b[cen_pos], ///
    by(schoolid) saving(ols, replace): reg z_read cen_pos
```

Now we open the 'ols' data set and produce some descriptive statistics. You could also examine some graphs if you prefer.

```
use ols, clear
su intercept coef
```

```
  Variable |  Obs              Mean      Std. Dev.             Min              Max
-----------+-----------------------------------------------------------------------
 intercept |  356         -.0343052        .5429935       -5.519699          1.79278
     coeff |  356           .013231        .0123741        -.034538         .1181929
```

```
sort coeff
```

```
list schoolid intercept coeff in 1/5
```

```
      +-------------------------------------+
      | schoolid    intercept          coeff |
      |-------------------------------------|
  1.  |      113    -.0731266       -.034538 |
  2.  |      234     .5434765      -.0281055 |
  3.  |       86     .5203956      -.0216063 |
  4.  |      195     -.137022      -.0203666 |
  5.  |      189    -.2271902       -.015418 |
      +-------------------------------------+
```

```
list schoolid intercept coeff in -5/-1
```

```
       +-------------------------------------+
       | schoolid    intercept          coeff |
       |-------------------------------------|
 352.  |       94    -.1233189       .0400667 |
 353.  |      297     .0791533       .0436432 |
 354.  |      153    -.0202645        .044675 |
 355.  |      221    -1.058333       .0823795 |
 356.  |      172      1.79278       .1181929 |
       +-------------------------------------+
```

We can see that the coefficients do vary quite a bit, so we will merge the 'ols' data set back into our chapter data and plot the regression lines for the first 12 schools.

```
use "data_for_chapter_3",clear
merge m:1 schoolid using "ols"

gen yhat = intercept+(coeff*cen_pos)
sort schoolid cen_pos
line yhat cen_pos if schoolid<13, connect(ascending) xline(0)  yline(0)
```

The graph in Figure 3.2, taken with the descriptive statistics, provides prima facie evidence that forcing a single coefficient (parallel slopes) for each school regression may not be the most appropriate way of modeling the relationship between parental occupational status and reading scores, given the large differences between some of the schools.

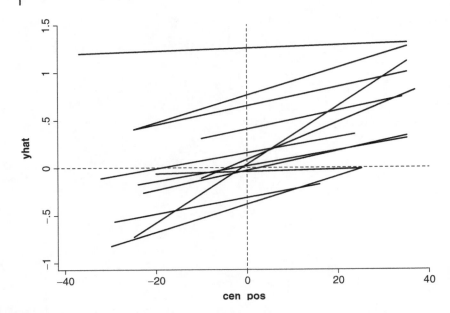

Figure 3.2 Individual regression lines of centred parental occupational status on standardized reading scores (first 12 schools by schoolid)

So, going back to the **mixed** command, we model a random coefficient by indicating the desired variable (*cen_pos*) after the colon (:) on the right-hand side of the command:

```
mixed z_read cen_pos || schoolid: cen_pos, ///
       stddev covariance(unstructured) nolog
```

We must also request an unstructured covariance matrix in the options part of the command – **covariance(unstructured)** or **cov(un)** – as the default is that the covariance (and therefore the correlation) is set to zero. It is important to do this because we want to allow the random intercepts and coefficients to covary with one another. As we will show shortly, the correlation between the random intercepts and coefficients is something that we want to investigate; so fixing this correlation to zero would result in losing some useful information. Always remember to override this default.

The output from this command is shown in Table 3.1. In the top part of the output, the value of the coefficient for *cen_pos* is 0.014 and the constant is –0.027. So, for the overall sample, we predict an increase of 0.014 in standardized reading scores for each increase in a unit of parental occupational status. The expected

Table 3.1 Stata output for random coefficient model of reading scores regressed on parental occupational status, with school grouping

```
Mixed-effects ML regression              Number of obs      =        13646
Group variable: schoolid                 Number of groups   =          356

                                         Obs per group: min =            3
                                                        avg =         38.3
                                                        max =           57

                                         Wald chi2(1)       =       742.96
Log likelihood = -17658.432              Prob > chi2        =       0.0000

------------------------------------------------------------------------------
      z_read |      Coef.   Std. Err.      z    P>|z|     [95% Conf. Interval]
-------------+----------------------------------------------------------------
     cen_pos |   .0135258   .0004962    27.26   0.000     .0125532    .0144983
       _cons |  -.0268864   .0252714    -1.06   0.287    -.0764176    .0226447
------------------------------------------------------------------------------


------------------------------------------------------------------------------
  Random-effects Parameters  |   Estimate   Std. Err.   [95% Conf. Interval]
-----------------------------+------------------------------------------------
schoolid: Unstructured       |
              sd(cen_pos)    |   .001396    .002252     .0000591    .0329635
              sd(_cons)      |   .4527814   .0200694    .4151062    .493876
      corr(cen_pos,_cons)    |  -.435798    .7759143   -.9817698    .8875512
-----------------------------+------------------------------------------------
              sd(Residual)   |   .8549859   .0053141    .8446337    .865465
------------------------------------------------------------------------------
LR test vs. linear regression:        chi2(3) =  1892.83   Prob > chi2 = 0.0000
```

standardized reading score for a student with mean parental occupational status (i.e. *cen_pos* = 0) is –0.027.

In the random effects parameters section of the output, there are three parts associated with 'schoolid: Unstructured'. The first two are the random parts of the random coefficients equation, one for the independent variable and one for the intercept:

- sd(cen_pos) is the standard deviation of the coefficients of the *cen_pos* variable, or a measure of how much the slopes vary across schools;
- sd(_cons) is the random intercept standard deviation as in the random intercept models (Level 2).

The last part in this section of the output is:

- corr(cen_pos,_cons), which is the correlation between the random parts of the coefficient and the intercept. The value here is negative, which means that schools with higher intercepts (higher reading scores for students with average parental occupational status) have smaller coefficients for the independent variable (smaller increases, or even decreases, in reading scores for each unit of parental occupational status). We can say this because we have centred the variable for parental occupational status so it has a 'meaningful mean'. If you choose to use variables that are not mean centred then the interpretation may not be straightforward. The interpretation of the correlation between the random intercepts and coefficients also needs the context of the overall direction of the regression lines, and so can be a bit confusing at first. We shall return to this in a moment in the section on 'Fanning in and fanning out'.

The last box of the output is:

- sd(Residual), which is the standard deviation at the student level (Level 1) as in the random intercepts output.

As we specified **stddev** as an option in the **mixed** command, the standard deviations of the coefficient, sd(cen_pos), and the intercept, sd(_cons), are reported. This also results in the reporting of the correlation between the random coefficients and intercepts, corr(cen_pos,_cons), rather than the covariance. You can easily obtain the covariance matrix by using a post-estimation command **estat recovariance**. As correlations are intuitively easier to understand, we focus here on the correlation between the random coefficients and intercepts.

At the bottom of the output you will see:

```
LR test vs. linear regression:chi2(3) =1892.83 Prob > chi2 = 0.0000
```

This test compares the current random coefficient model to the single-level model. The chi-square (1892.83) is statistically significant, but this is not a surprise, as we already knew from the previous chapter that it is important to account for the different school intercepts. This particular likelihood ratio test does not help us assess whether adding the random coefficient was a good thing or not. So, is this a good model? We need to test if adding the random coefficient component improved the model fit. In order to test this we just need to save the results of our random coefficient model:

```
estimates store rc
```

We then run the model without the random coefficient, store the results, and perform a likelihood ratio test between these two models.

```
quietly mixed z_read cen_pos || schoolid:, nolog

estimates store ri

lrtest rc ri
```

Likelihood-ratio test   LR chi2(2) =   1.24
(Assumption: ri nested in rc)   Prob > chi2 = 0.5391

Note: The reported degrees of freedom assumes the null hypothesis is not
  on the boundary of the parameter space. If this is not true, then the
  reported test is conservative.

We can now see that the addition of the random coefficient did not improve the model, and so we would conclude that is not worth pursuing the random coefficient modeling of parental occupational status. In this case, Stata has produced a note for the p value, but there is no need to get tied up with it at this stage.

It is important to note that you should always check whether or not the random coefficient is worth pursuing statistically, because you do not want to add unnecessary degrees of complexity to your model. You should also not add unnecessary complexity to your model just because you can – your empirical models should always be closely tied to a theoretical/conceptual model. We have emphasized this point in the previous chapter as well, but we think it is worth repeating because there is a tendency among new users to try to do everything in one model, only to create large, complex estimations that are very difficult to interpret.

TRYING A DIFFERENT RANDOM COEFFICIENT

Our example with parental occupational status resulted in the conclusion that it was best modeled as a 'fixed' coefficient. However, we would like to demonstrate how a random coefficient may be modeled, so we will use PISA's index of economic, social and cultural status (*escs*) in the following examples. This is a composite index variable that was created by items from the student questionnaire, including parental occupation and education as well as possessions within the home (e.g. books, computers, areas for children to study). We refer to this as 'family capital' for short. The variable has been centred (*cen_escs*) so that zero represents the sample mean, as we have done in previous examples. It might be the case that you are investigating a theory that family capital differentially affects reading scores depending on the school that the child attends.

We will test to see if our random coefficient model is an improvement on the random intercept model.

We start with the random intercept model:

```
mixed z_read cen_escs || schoolid:, stddev nolog

estimates store ri
```

Table 3.2   Stata output for random intercept model of reading scores regressed on family capital, with school grouping

```
Mixed-effects ML regression              Number of obs      =         13644
Group variable: schoolid                 Number of groups   =           356

                                         Obs per group: min =             3
                                                        avg =          38.3
                                                        max =            57

                                         Wald chi2(1)       =       1101.50
Log likelihood = -17501.482              Prob > chi2        =        0.0000


----------------------------------------------------------------------------
      z_read |    Coef.    Std. Err.     z     P>|z|    [95% Conf. Interval]
-------------+--------------------------------------------------------------
    cen_escs |   .355476   .0107107   33.19   0.000    .3344834    .3764686
       _cons | -.0234824   .0233226   -1.01   0.314   -.0691938    .0222289
----------------------------------------------------------------------------


----------------------------------------------------------------------------
  Random-effects Parameters   |  Estimate   Std. Err.   [95% Conf. Interval]
------------------------------+---------------------------------------------
schoolid: Identity            |
                 sd(_cons)    |  .4145822    .01881     .3793069    .4531381
------------------------------+---------------------------------------------
              sd(Residual)    |  .8471417   .0052072    .836997     .8574094
----------------------------------------------------------------------------
LR test vs. linear regression: chibar2(01) = 1562.60 Prob >= chibar2 = 0.0000
```

We can see from the output in Table 3.2 that the random intercept model is an improvement over the single-level model. So we proceed to adding the random coefficient:

```
mixed z_read cen_escs || schoolid: cen_escs, stddev cov(un) nolog

estimates store rc

lrtest rc ri
```

The likelihood ratio test shown in Table 3.3 is statistically significant. We therefore have evidence that modeling family capital as a random coefficient is appropriate.

Table 3.3 Stata output for random coefficient model of reading scores regressed on family capital, with school grouping

```
Mixed-effects ML regression              Number of obs      =       13644
Group variable: schoolid                 Number of groups   =         356

                                         Obs per group: min =           3
                                                        avg =        38.3
                                                        max =          57

                                         Wald chi2(1)       =      730.04
Log likelihood = -17476.482              Prob > chi2        =      0.0000

------------------------------------------------------------------------------
     z_read |      Coef.   Std. Err.      z    P>|z|     [95% Conf. Interval]
------------+-----------------------------------------------------------------
   cen_escs |   .3548524   .0131333    27.02   0.000     .3291116    .3805933
      _cons |  -.0150228   .0230328    -0.65   0.514    -.0601663    .0301206
------------------------------------------------------------------------------


------------------------------------------------------------------------------
  Random-effects Parameters  |   Estimate   Std. Err.   [95% Conf. Interval]
-----------------------------+------------------------------------------------
schoolid: Unstructured       |
               sd(cen_escs)  |   .1395733   .0161092    .1113163    .1750032
                  sd(_cons)  |    .407023   .0185398    .3722605    .4450317
         corr(cen_escs,_cons)| -.3972283   .0956085   -.5668374   -.1953145
-----------------------------+------------------------------------------------
               sd(Residual)  |   .8420302   .0052293    .8318431    .8523421
------------------------------------------------------------------------------
LR test vs. linear regression:      chi2(3) = 1612.60  Prob > chi2 = 0.0000

Note: LR test is conservative and provided only for reference.

. estimates store rc

. lrtest rc ri

Likelihood-ratio test                            LR chi2(2)  =    50.00
(Assumption: ri nested in rc)                    Prob > chi2 =   0.0000

Note: The reported degrees of freedom assumes the null hypothesis is not on
the boundary of the parameter space. If this is not true, then the reported
test is conservative.
```

The value of the coefficient for *cen_escs* is 0.355 and the constant is –0.015. Therefore, in the overall sample there is a predicted 0.355 increase in standardized reading scores for each unit increase in family capital. From the intercept we expect a standardized reading score of –0.015 for a student with the mean level of family capital (*cen_escs* = 0) and in an 'average' school. What we mean here by an 'average' school is one that has the same intercept as the overall sample, that is, where the random part of the intercept is equal to zero so there is no deviation from the overall sample intercept. It may be the case that none of the schools actually have that value.

Looking at the random parts of the estimation:

- sd(cen_escs) is the standard deviation of the random coefficient of the *cen_escs* variable or, in other words, the standard deviation of the deviations of the school-specific coefficients (slopes) from the overall sample coefficient.
- sd(_cons) is the standard deviation of the intercept (i.e. when *cen_escs* = 0) or Level 2 residuals and the same as in the random intercept models.
- corr(cen_escs,_cons) is the correlation between the random intercepts and the random coefficients.

To interpret the correlation between the random intercepts and the random coefficients, here a value of –0.397, we first look at a scatterplot of the intercepts and coefficients:

```
*first, generate predicted random effects for the slope and intercept
predict u1 u0, reffects
*second, pick one observation per school
egen pickone=tag(schoolid)
*third, create a scatterplot with each dot representing a school
scatter u1 u0 if pickone==1, yline(0) xline(0)
```

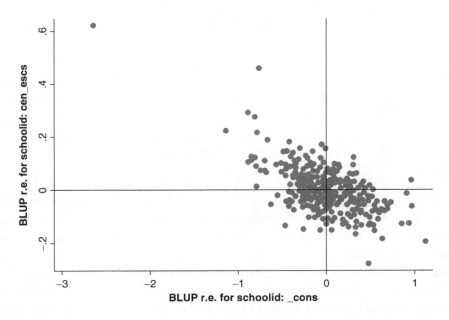

Figure 3.3 Scatterplot of random intercepts and random coefficients

In Figure 3.3 each dot represents a school with its random component of the intercept ($u0$) on the x-axis and the random component of the coefficient ($u1$) on the y-axis. Remember that these values are the differences for each school from the overall sample intercept and coefficient.

Therefore, all schools to the left of zero on the horizontal axis (vertical line) are schools with lower average reading scores than the overall sample average reading score for students with mean family capital (i.e. *cen_escs* = 0). And those schools to the right of zero are those with higher reading scores than the overall sample for students with mean family capital. For the coefficients for family capital, those schools that are above zero on the vertical axis (horizontal line) have a larger coefficient, while those that are below zero have a smaller coefficient.

If you prefer you can use the actual values for each school's intercept and coefficient rather than their differences from the overall sample intercept and coefficient by adding $u0$ and $u1$ to the values from the output. Examining the descriptive statistics of these new variables may also be useful as well as plotting them.

```
gen intercept=-0.015+u0 if pickone==1
gen coeff=0.3548+u1 if pickone==1
su intercept coeff if pickone==1
```

Table 3.4 Descriptive statistics for group intercepts and coefficients

Variable	Obs	Mean	Std. Dev.	Min	Max
intercept	356	-.015	.3819469	-2.658031	1.113941
coeff	356	.3548	.0867643	.0757405	.9771496

We can see from the descriptive statistics of the new variables – *intercept* and *coeff* – that actual values for the intercept range between –2.66 and 1.11 and the coefficients for family capital range from 0.076 to 0.997 (Table 3.4). From this we can see that for all schools the value of the family capital coefficient (or slope) is positive. In the scatterplot in Figure 3.4 we have used crosses for each school and put the lines at the actual values for the overall sample intercept and coefficient.

```
scatter coeff intercept, msymbol(x) msize(large) ///
       yline(0.3548) xline(-0.015)
```

The overall direction of the plot is from upper left to lower right. This is because the correlation between the two is negative, as seen in the output in Table 3.3.

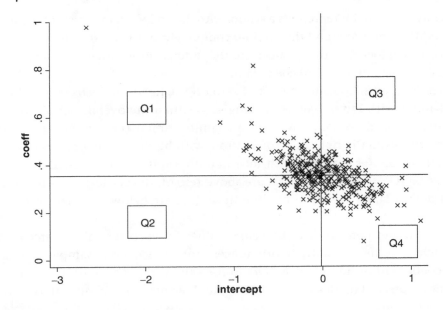

Figure 3.4 Scatterplot of group intercepts and coefficients

If the correlation were positive, the general direction of the plots would be from lower left to upper right.

Now, let us look at the graph by quadrant, starting with the top left quadrant (Q1). Schools in this quadrant have below average reading scores for students with mean family capital (i.e. lower intercepts), but larger than average coefficients for family capital (i.e. family capital has a larger effect on reading scores). Below them in the lower left quadrant (Q2) are schools with below average reading scores for students with mean family capital but the coefficients are smaller than average (i.e. smaller effects for family capital). In the upper right quadrant (Q3) are schools with above average reading scores for students with mean family capital along with larger coefficients for family capital. Below them in the lower right quadrant (Q4) are schools with above average reading scores for students with mean family capital but with smaller coefficients for family capital.

Most of the schools are either in Q1 or Q4, which is reflected in the moderate negative correlation reported in the output (Table 3.3). This generally indicates that in schools with lower average reading scores there is a larger effect of family capital and in schools with higher average reading scores there is a smaller effect of family capital. Clearly this doesn't hold for all schools as there are schools in Q2 and Q4, but the overall trend is a negative correlation.

SHRINKAGE

In Figure 3.5, the graph on the left shows the predicted individual schools lines from the random coefficient model. One the right, for purposes of contrast, is the graph created by individual OLS regressions (one per school). You can see that they are similar, but that the graph from the random coefficient model is more tightly bundled. This is because the estimation procedure creates a *common* slope and intercept as well as the random terms for each individual school. You are seeing the phenomenon of 'shrinkage' wherein the intercepts' distance from the overall mean line 'shrinks' because multilevel modeling techniques use information from all the groups (i.e. schools) in the estimate for each individual group. You see more 'shrinkage' when Level 1 variance is large and Level 2 variance is small. At this stage we will leave the discussion of shrinkage at this basic level but we recommend that, when you are more familiar with multilevel models, you investigate the matter further; for example, Tate (2004) is quite accessible.

```
predict predscore, fitted
line predscore cen_escs, connect(ascending) ///
      title(Random Coefficient Regression Lines)
graph save rc, replace


statsby sa=_b[_cons] sb=_b[cen_escs], by(schoolid) ///
      saving(ols, replace): regress z_read cen_escs
merge m:1 using ols
drop _merge
gen yhat = sa+sb*cen_escs
sort schoolid cen_escs
line yhat cen_escs, connect(ascending) ///
   title(OLS Regression Lines)
graph save ols, replace


graph combine rc.gph ols.gph, ycommon
```

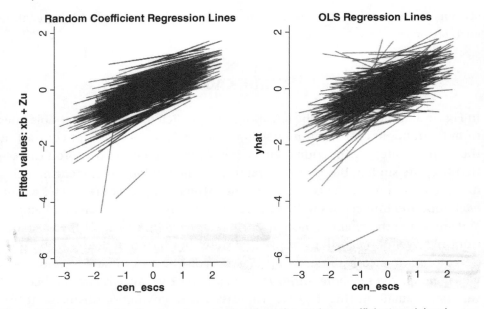

Figure 3.5 Comparison of group regression lines from the random coefficient model and separate OLS regressions

FANNING IN AND FANNING OUT

When the correlation between the random slopes and intercepts is negative, the graphical depiction of the correlation will be one of 'fanning in' of the regression lines (Figure 3.6). This is what we observe in our current example. In other words, there are more pronounced differences between schools at low levels of family capital than at higher levels of family capital. So, in our example with all school regression lines having a positive slope it can be seen in the left-hand graph of Figure 3.5 that at higher values of family capital (right side of the graph), the lines converge. What makes the lines fan in is that generally those lines with the lower intercepts have steeper slopes (larger coefficients). At the top of the graph, we can see that regression lines that have higher intercepts have flatter trajectories (smaller coefficients). In general, these regression lines from the random coefficient model are telling us that family capital has a stronger effect on pupils who are in schools with lower than average reading scores in the same way we have interpreted the scatterplot in Figure 3.4.

If the correlation between the random slopes and intercepts had been positive, the graphical depiction of the correlation would be one of 'fanning out' of the regression lines (Figure 3.7).

If the correlation were close to zero, then there would be no fanning of the regression lines in the graph – just a scattering of lines (Figure 3.8).

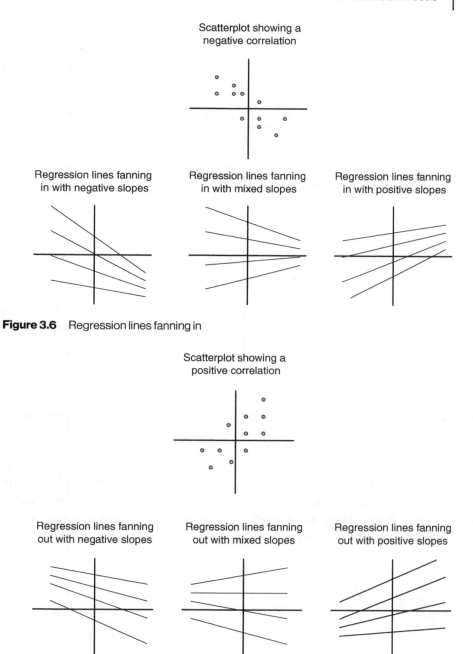

Figure 3.6 Regression lines fanning in

Figure 3.7 Regression lines fanning out

To interpret the correlation and graphs it is useful to know the direction of the slopes for all or most of your groups. In our example we can see from the descriptive statistics in Table 3.4 that all school regression lines are positive.

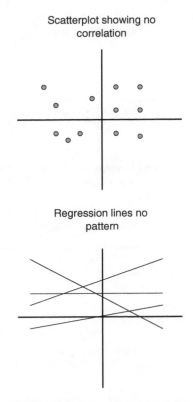

Scatterplot showing no
correlation

Regression lines no
pattern

Figure 3.8 No fanning of regression lines because the correlation between random intercepts and coefficients is zero

Another way to determine the direction and spread of the regression lines is to estimate the 'predictive' intervals around the random coefficients (Rabe-Hesketh and Skrondal, 2008). Here is the part of the output of the random coefficient model with standard deviations from Table 3.3:

```
------------------------------------------------------------------------------
      z_read |   Coef.      Std. Err.  z   P>|z|   [95% Conf. Interval]
------------------------------------------------------------------------------
    cen_escs |  .3548524  .013133327 .02  0.000    .3291116     .3805933
       _cons | -.0150228  .0230328-0 .65  0.514   -.0601663     .0301206
------------------------------------------------------------------------------


------------------------------------------------------------------------------
 Random-effects Parameters |  Estimate  Std. Err.  [95% Conf. Interval]
---------------------------+--------------------------------------------------
schoolid: Unstructured     |
          sd(cen_escs)     |  .1395733  .0161092   .1113163     .1750032
```

```
            sd(_cons) |  .407023    .0185398   .3722605   .4450317
  corr(cen_escs,_cons) | -.3972283   .0956085  -.5668374  -.1953145
----------------------+-----------------------------------------------
          sd(Residual) |  .8420302   .0052293   .8318431   .8523421
----------------------+-----------------------------------------------
```

Assuming that the distribution of the random effects is normal (we test this later in the chapter in the section on diagnostics), the 95% predictive interval is ±1.96 standard deviations from the mean. From this output, the 'fixed' effect for family capital (*cen_escs*) is the mean effect and the standard deviation of all the random coefficients from that fixed effect is shown at sd(cen_escs). So, the 95% predictive interval for the random coefficient for family capital is:

$$0.3549 \pm (1.96 \times 0.1396) = [0.081, 0.629]$$

From this, we would expect that 95% of schools have coefficients between 0.081 and 0.629. As noted by Rabe-Hesketh and Skrondal (2008), this exercise of creating 'predictive' intervals for random coefficients is particularly useful if we have reason to believe that some schools have coefficients that are positive and others are negative. This is not the case here, however, and all coefficients are positive, which can be seen in the post-estimation graphs shown earlier.

EXAMINING THE VARIANCES

Next we move on to examining the variances for the models. By having a random coefficient model for family capital, we are allowing the between-school variance to differ across various levels of family capital. But we need to test if there actually is any difference in the variances.

First we look at the Level 1 and Level 2 variance for the random intercept model and compare it to the random coefficient model. From some quick calculations from the outputs in Tables 3.2 and 3.3 to get variances instead of standard deviations we have the following:

	Random intercept (Table 3.2)	Random coefficient (Table 3.3)
School-level variance (Level 2)	0.172	0.166
Student-level variance (Level 1)	0.718	0.709

We see that the Level 2 variance reduces slightly from 0.172 to 0.166, while the Level 1 variance goes from 0.718 to 0.709. The Level 1 residual is reduced because the non-parallel lines allowed by the random coefficient model make for a better-fitting model.

In the random intercept model, the Level 2 variance is constant across values of family capital as the regression lines are parallel to each other. When we add the random coefficient to the model, the Level 2 variance is conditional on the value of family capital, the independent variable. It may be that there are no significant differences in the Level 2 variance across values of the independent variable, but, with the regression lines in our example fanning in, it is likely that the Level 2 variance will change with values of the independent variable. Therefore, it is a little more complex to calculate. In words, the Level 2 variance with one coefficient is equal to:

variance of the intercepts

> + [2 × (covariance of intercepts and coefficients) × x]

> + [variance of the coefficients × x^2]

From the model in Table 3.3, below is the random-effects section of the output but with variances instead of standard deviations:

```
------------------------------------------------------------------------
Random-effects Parameters    |    Estimate   Std. Err.   [95% Conf. Interval]
-----------------------------+------------------------------------------
schoolid: Unstructured       |
             var(cen_escs)   |    .0194807   .0044968    .0123913    .0306261
               var(_cons)    |    .1656677   .0150922    .1385778    .1980532
      cov(cen_escs,_cons)    |   -.0225664   .0058796   -.0340902   -.0110425
-----------------------------+------------------------------------------
             var(Residual)   |    .7090149   .0088065    .6919629    .7264871
------------------------------------------------------------------------
```

Using some of the notation from the output, the Level 2 variance is equal to:

$$\text{var(\_cons)} + [2 \text{ cov(cen\_escs, \_cons)} \, x] + [\text{var(cen\_escs)} \, x^2]$$

where x is equal to the values of the *cen_escs* variable. Substituting the values from the output:

$$0.1657 + [2 \times (-0.0226) \times x] + [0.0195 \times x^2]$$

You can graph this function by using the **twoway function** command and then giving plausible values of the covariate of interest (in this case, *cen_escs* with a range from −4 to 2). This graph does not actually use any of the data in memory but instead draws the line associated with the function that we specify.

```
twoway function 0.1657 + [2*-0.0226*x] + [0.0195*x^2], ///

    range(-4 2) xtitle(cen_escs) ///

    ytitle(Estimates of between-school variance)
```

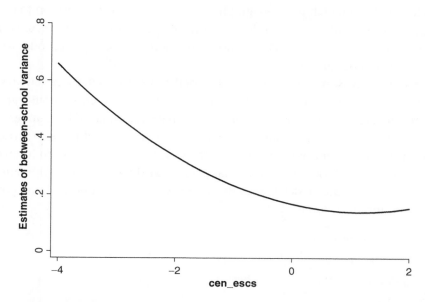

Figure 3.9 Graph of between-school variance against family capital

The resulting graph (Figure 3.9) shows that there is more variance between schools at the lower levels of family capital and less variance when family capital is high. This is another way of showing the result of the school-specific regression lines fanning in and thus the differences in variance across the schools at different levels of family capital.

In Chapter 2 we showed how, for random intercept models, the ICC/rho could be calculated from the variance of the random intercept and the variance of the Level 1 residual. As we have noted above, in a random coefficient model the Level 2 variance is not constant as the regression lines are not parallel to each other, and so is conditional upon the value of the independent variable. As the Level 2 variance (variance of the random intercept) is a component of the ICC/rho it follows that the ICC/rho is also not constant and is conditional on the value of the independent variable. The ICC/rho value given in the Stata output after the **xtmrho** command is when the independent variables have a value of zero. This is another good reason for centring the independent variables so that the ICC/rho is meaningful (Rabe-Hesketh and Skrondal, 2008).

A DICHOTOMOUS VARIABLE AS A RANDOM COEFFICIENT

So far we have used continuous variables as independent variables in our random coefficient models. Continuous variables are generally the most straightforward to model and from which to create graphs. It is, however, quite common to model nominal-level variables with random coefficients, and the following examples will demonstrate how to do this. We start with a random coefficient model with a single independent variable.

We may have reason to believe that the coefficients for males and females are different by *schoolid*. If so, we can test this by adding the dichotomous variable *female* to the fixed and random portions of the **mixed** command. Remember, usually you want to put the random coefficient in the fixed part of the estimation as well (i.e. not only as random). First we run the model with only random intercepts and then do a likelihood ratio test to make sure that the inclusion of the random coefficient improves the model. We use the **quietly** prefix to suppress the output for the random intercept model.

```
quietly mixed z_read female || schoolid:

estimates store ri

mixed z_read female || schoolid:female, cov(un) nolog

estimates store rc

lrtest rc ri
```

In the main output in Table 3.5, the fixed part of the estimation shows that the coefficient for *female* is 0.407, meaning that in the overall sample female students have, on average, higher standardized reading scores by 0.407. The constant is –0.240, which is the overall sample mean standardized reading score for males (i.e. when *female* = 0).

Turning to the random effects, the variance of the coefficient, var(female), is 0.024 and the variance of the intercept, var(_cons), is 0.294. The covariance of the random slopes and intercepts is –0.035. We can generate the correlation instead by using the command:

```
estat recov, corr

Random-effects correlation matrix for level schoolid

              |      female                        _cons
--------------+-------------------------------------------
       female |          1
        _cons |   -.4133897                           1
```

Table 3.5 Stata output for random coefficient model of reading scores regressed on gender, with school grouping

```
Mixed-effects ML regression              Number of obs      =       13646
Group variable: schoolid                 Number of groups   =         356

                                         Obs per group: min =           3
                                                        avg =        38.3
                                                        max =          57

                                         Wald chi2(1)       =      468.86
Log likelihood = -17714.728              Prob > chi2        =      0.0000

------------------------------------------------------------------------------
     z_read |      Coef.   Std. Err.     z    P>|z|     [95% Conf. Interval]
------------+-----------------------------------------------------------------
     female |   .4067323    .018784   21.65   0.000     .3699164    .4435482
      _cons |  -.2402186   .0310616   -7.73   0.000    -.3010982    -.179339
------------------------------------------------------------------------------


------------------------------------------------------------------------------
  Random-effects Parameters  |   Estimate   Std. Err.    [95% Conf. Interval]
-----------------------------+------------------------------------------------
schoolid: Unstructured       |
               var(female)   |   .0241164   .0087197     .0118726    .0489866
                var(_cons)   |   .2941922   .0268464     .2460113    .3518092
        cov(female,_cons)    |  -.0348202   .0127956    -.0598991   -.0097414
-----------------------------+------------------------------------------------
             var(Residual)   |   .7289715   .0090504     .7114472    .7469275
------------------------------------------------------------------------------
LR test vs. linear regression:      chi2(3) =  2802.86   Prob > chi2 = 0.0000

Note: LR test is conservative and provided only for reference.

. estimates store rc

. lrtest rc ri

Likelihood-ratio test                          LR chi2(2)   =     15.06
(Assumption: ri nested in rc)                  Prob > chi2  =    0.0005
```

This moderate negative correlation (–0.413) means that there is evidence of 'fanning in'. In terms of the dichotomous variable *female*, this means that the positive effect on reading scores of being female is larger in schools with lower average reading scores for males (i.e. the intercept when *female* = 0).

The likelihood ratio test is statistically significant, indicating that we have evidence that the effect of gender varies across schools, with a better-fitting model.

You could graph these effects as we did in the earlier example for family capital, first with individual regression lines and then with the regression lines produced by the random coefficient model. The commands for these graphs are identical to the ones previously shown, except for the variable names that are specific to this example. However, as *female* is a dichotomous variable, the line graphs would only be illustrating the difference between two categories – being male (*female* = 0) and being female (*female* = 1). These graphs are probably less informative than the previous example, and for a dichotomous random coefficient descriptive statistics would probably be enough to see what is going on in the model:

```
predict u1 u0, reffects

egen pickone=tag(schoolid)

su u1 u0 if pickone==1

    Variable |    Obs       Mean    Std. Dev.        Min        Max
-------------+------------------------------------------------------
          u1 |    356   9.29e-11    .0763284  -.2419131   .2710053
          u0 |    356   3.56e-10    .5168739  -3.895441   1.553343
```

We can see that the random component of the coefficient (*u1*) ranges from 0.242 below the overall sample coefficient (0.407, Table 3.5) to 0.271 above. Thus, for all schools the coefficient for *female* is positive.

In this model, where we only have one dichotomous independent variable, the intercepts are the mean reading score for males in each school. There is considerable range in the intercepts compared to the range in the coefficients, which also indicates fanning in. Returning to our equation for the variance between schools for values of the independent variable,

$$var(\_cons) + [2\ cov(female, \_cons)\ x] + [var(female)\ x^2]$$

where x is the variable *female*, we have

$$0.2942 + [2 \times (-0.0348) \times x] + [0.0241 \times x^2]$$

We don't need Stata to graph this as the arithmetic is easy with a zero–one variable to substitute for x. The variance across school for females is $0.2942 + [2 \times (-0.0348)] + 0.0241 = 0.2487$ and for males is 0.2942, which shows more variation in reading scores for males between schools than for females – again, evidence of fanning in.

MORE THAN ONE RANDOM COEFFICIENT

So far we know that a model with a random coefficient for *cen_escs* is better than a model without the random coefficient and that, in another model, a random coefficient for *female* is a better model. But could it be the case that a model with both random coefficients would be even better?

You can have numerous random coefficients in a model. The problem is that the model soon becomes very complicated as each additional random coefficient adds parameters to it and new lines of output to read and make sense of. Compare the output in Tables 3.5 and 3.6 for just one additional random coefficient.

First, let us see if the model with two random coefficients is worth pursuing:

```
quietly mixed z_read cen_escs female || ///
        schoolid: cen_escs, cov(un) var
estimates store rc1

quietly mixed z_read cen_escs female || ///
        schoolid: cen_escs female, cov(un) var
estimates store rc2

lrtest rc1 rc2

Likelihood-ratio test                 LR chi2(3)  =     20.93
(Assumption: rc1 nested in rc2)       Prob > chi2 =    0.0001
```

The likelihood ratio test is statistically significant, indicating that the inclusion of the second random coefficient improves the model fit. So, let's look at the output for the model with two random coefficients.

```
estimates replay rc2
```

The fixed part of the model is fairly straightforward, with the fixed effects for *cen_escs*, *female* and the *constant* reported (Table 3.6). The fixed coefficients for the two independent variables are very similar to those reported in our bivariate examples earlier in this chapter (Tables 3.3 and 3.5). The interpretations are similar to those before, except that the effect of each independent variable is a net effect controlling for the other independent variable. The intercept is the reading score for male students (*female* = 0) with mean family capital (*cen_escs* = 0) in an 'average' school.

The random effects, however, now consist of six parts. You might find that having more than one random coefficient results in Stata taking more time to converge as well. The variances for the random coefficients and intercepts are displayed first, with var(cen_escs) = 0.017, var(female) = 0.022, and var(_cons) = 0.186. The values for the two independent variables, *cen_escs* and *female*, represent the Level 2 (between-school) variance in the coefficients of these variables. The intercept variance is the between-school variance when both *cen_escs* and *female*

Table 3.6 Stata output for random coefficient model of reading scores regressed on family capital and gender, with school grouping

```
-------------------------------------------------------------------------------
Model rc2
-------------------------------------------------------------------------------

Mixed-effects ML regression                 Number of obs        =        13644
Group variable: schoolid                    Number of groups     =          356

                                            Obs per group: min   =            3
                                                           avg   =         38.3
                                                           max   =           57

                                            Wald chi2(2)         =      1388.29
Log likelihood = -17142.875                 Prob > chi2          =       0.0000


------------------------------------------------------------------------------
      z_read |      Coef.   Std. Err.      z    P>|z|     [95% Conf. Interval]
-------------+----------------------------------------------------------------
    cen_escs |    .3544379   .0126621    27.99   0.000     .3296207    .3792551
      female |    .4039403   .0179278    22.53   0.000     .3688023    .4390782
       _cons |   -.2162411    .025544    -8.47   0.000    -.2663065   -.1661758
------------------------------------------------------------------------------


------------------------------------------------------------------------------
  Random-effects Parameters  |   Estimate   Std. Err.    [95% Conf. Interval]
-----------------------------+------------------------------------------------
schoolid: Unstructured       |
              var(cen_escs)  |   .0173153   .0041766     .0107923    .0277808
               var(female)   |   .0219747   .0080396     .0107276    .0450135
                var(_cons)   |   .1859746    .018297     .1533588    .2255271
      cov(cen_escs,female)   |   -.005392   .0041496    -.0135251    .0027411
       cov(cen_escs,_cons)   |  -.0219697   .0061918    -.0341054    -.009834
       cov(female,_cons)     |  -.0306238   .0099086    -.0500443   -.0112034
-----------------------------+------------------------------------------------
              var(Residual)  |   .6721244   .0084257     .6558116    .6888429
------------------------------------------------------------------------------
LR test vs. linear regression:      chi2(6) = 1685.56   Prob > chi2 = 0.0000
```

are equal to zero (i.e. males with mean family capital). And then there are the three covariances reported. We can use the command **estat recov, corr** to have the covariances presented as correlations, which may be easier to interpret.

```
Random-effects correlation matrix for level schoolid

             |    cen_escs      female       _cons
-------------+---------------------------------------
    cen_escs |          1
      female |   -.2764222           1
       _cons |   -.3871529   -.4790399           1
```

The correlation between the random coefficient for *cen_escs* and the random intercept is –0.387, suggesting fanning in, and the correlation between the random coefficient for *female* and the random intercept is –0.479, which suggests similar (but a somewhat stronger pattern of) fanning in. These correlations are very similar to what we saw when we ran the separate models earlier.

There is a new correlation reported between the random coefficients for the two independent variables, *female* and *cen_escs*, which is negative and weak at –0.276. As it is a weak correlation it is probably not very interesting, but to follow through on the full process in our example, we can generate a scatterplot between the random components of the two independent variables in a similar way as earlier in this chapter.

```
predict u1 u2 u0, reffects
scatter u1 u2 if pickone==1, yline(0) xline(0)
```

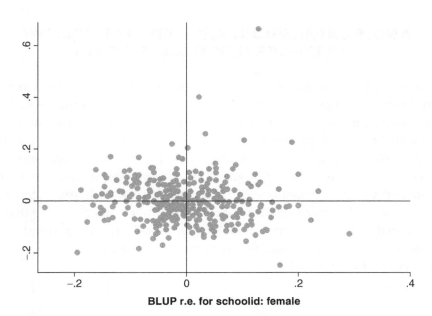

BLUP r.e. for schoolid: female

Figure 3.10 Scatterplot of random coefficients for family capital and gender

The resulting scatterplot is shown in Figure 3.10. Remember that these plots are relative to the overall sample coefficients reported in the upper part of the output. We can work out that the coefficients for *female* and *cen_escs* are positive for all schools. If you wish to see this you can run the same commands as we did earlier and display descriptive statistics.

There is a very slight left to right downward trajectory in the scatterplot, which we would expect given the weak negative correlation reported in the output. From the graph alone, this correlation is difficult to determine. Therefore, the effect of family capital is stronger (larger coefficients) when there are smaller effects from being female (top left quadrant). Or, alternatively, there are smaller effects of family capital when there is a large gender effect (lower right quadrant).

These covariances and/or correlations between random coefficients are somewhat like 'interaction terms' in that they reveal how coefficients may vary across values of the coefficient for the other variable. This may not be intrinsically interesting to your research question, but they can be quite large and often tell an interesting story. These covariances and/or correlations are often not reported in published tables, but it is important to include them in your estimation procedure (through the unstructured covariance option that we have been using).

A NOTE ON PARSIMONY AND FITTING A MODEL WITH MULTIPLE RANDOM COEFFICIENTS

It is possible to fit models with multiple random coefficients as well as models with coefficients representing the categories of a nominal variable with more than two categories. In the previous example, the additional random coefficient added three new parameters to the model, so imagine how many would be added if you included a nominal variable with five categories. Your models can quickly become complex and the output rather cumbersome. As a new user, you are advised to use as parsimonious models as possible until you become more confident with the interpretations. Likelihood ratio tests may indicate that more complicated models are 'better' than simpler models, but these statistical tests do not tell you anything about the more meaningful and substantive aspects of interpreting of your model.

A MODEL WITH ONE RANDOM AND ONE FIXED COEFFICIENT

At the start of this chapter we continued our example from Chapter 2 where we investigated the effect of parental occupational status (*cen_pos*) on standardized reading scores. We found that using a random intercept model with schools as the grouping variable for parental occupational status was 'better' than a single-level model, but when we tried to model parental occupational status as a random coefficient there was no benefit in doing that. So, for our example we moved onto looking at another independent variable – family capital.

However, if we wanted to continue with our first example using parental occupational status and then add in gender, which our bivariate models suggest is better fitted using a random coefficient, this can be easily done within the **mixed** command:

```
mixed z_read cen_pos female || schoolid:female, ///
    stddev cov(un) nolog
```

Here we have the two independent variables, *cen_pos* and *female*, and after the fence (the two parallel vertical lines) we still specify *schoolid* as the first random component for the grouping variable intercepts, but only include *female* here to estimate the random coefficient. By putting *cen_pos* in the first part of the command but not in the second, after the fence, we are telling Stata to only estimate the 'fixed' effect of parental occupational status; that is, the effect is constrained to be the same across all schools.

Table 3.7 Stata output for random coefficient model of reading scores regressed on parental occupational status and gender, with school grouping

```
Mixed-effects ML regression              Number of obs      =       13646
Group variable: schoolid                 Number of groups   =         356

                                         Obs per group: min =           3
                                                        avg =        38.3
                                                        max =          57

                                         Wald chi2(2)       =     1289.95
Log likelihood = -17334.943              Prob > chi2        =      0.0000

------------------------------------------------------------------------------
     z_read |      Coef.   Std. Err.      z    P>|z|     [95% Conf. Interval]
------------+-----------------------------------------------------------------
    cen_pos |   .0134425   .0004787    28.08   0.000     .0125042    .0143807
     female |   .4058738   .0181235    22.39   0.000     .3703525    .4413952
      _cons |  -.2320921   .0276787    -8.39   0.000    -.2863413   -.1778429
------------------------------------------------------------------------------


------------------------------------------------------------------------------
  Random-effects Parameters |   Estimate   Std. Err.    [95% Conf. Interval]
----------------------------+-------------------------------------------------
schoolid: Unstructured      |
              sd(female)    |   .1458923   .0279528     .1002174    .2123839
              sd(_cons)     |   .4756953   .0227422      .433146    .5224244
       corr(female,_cons)   |  -.4323915   .1257504    -.6445841    -.158348
----------------------------+-------------------------------------------------
              sd(Residual)  |   .8327365   .0051737     .8226578    .8429387
------------------------------------------------------------------------------
LR test vs. linear regression:       chi2(3) =   1973.99  Prob > chi2 = 0.0000
```

This command produces the output shown in Table 3.7. We can see that for the variable *cen_pos* there is only a coefficient reported in the upper panel while for the variable *female* there is a 'fixed' coefficient in the upper panel along with the random effects parameters in the lower panel. We would interpret these elements in a similar way to before, except that the effect of parental occupational status is now controlling for gender and the effects of being female are controlling for parental occupational status.

ADDING LEVEL 2 VARIABLES

Level 2 variables are easily added to the models in the **mixed** command. In our examples, Level 2 variables are ones that measure characteristics of the schools and, therefore, are a constant for each individual school – all the students in that school have the same value for that variable. For example, if the size of the school is 1,000 students, then all students in that school will have the value 1,000 for the variable school size. For these Level 2 variables, as there is no variation within each school they can only be modeled as a 'fixed' effect in our two-level models and the interpretation can only be as an effect in the whole sample.

If we continue following our example, in Chapter 2 we went on to examine the effect of school size on reading scores. We created a centred variable for school size where school size is measured as the number of students. The resulting coefficient will be the predicted change in reading scores for each additional student – probably a level of detail we would not be interested in. So, we have created a new centred variable where the unit is per 100 students, *cen_size100*, to use in this model. We add the Level 2 variable to the fixed part of the command thus:

```
mixed z_read cen_pos female cen_size100 || ///
        schoolid: female, stddev cov(un) nolog
```

In the output in Table 3.8 we see that school size (*cen_size100*) has a significant 'fixed' effect but we should also test to see if the inclusion of the variable improves the model fit:

```
. lrtest model1 model2

Likelihood-ratio test                    LR chi2(1)  =  37.12
(Assumption: model1 nested in model2) Prob > chi2 = 0.0000
```

We saved the estimates from Table 3.7 as Model 1 and from Table 3.8 as Model 2. We can see that the likelihood ratio test indicates an improved fit, so this together with the significant effect would usually persuade us to retain this variable in our model.

Table 3.8 Stata output for random coefficient model of reading scores regressed on parental occupational status, gender, and school size, with school grouping

```
Mixed-effects ML regression            Number of obs      =       13646
Group variable: schoolid               Number of groups   =         356

                                       Obs per group: min =           3
                                                      avg =        38.3
                                                      max =          57

                                       Wald chi2(3)       =     1345.32
Log likelihood = -17316.384            Prob > chi2        =      0.0000

-------------------------------------------------------------------------
      z_read |      Coef.  Std. Err.      z    P>|z|   [95% Conf. Interval]
-------------+-----------------------------------------------------------
     cen_pos |   .0133631  .0004789   27.90   0.000    .0124244   .0143018
      female |   .4075346  .0180479   22.58   0.000    .3721614   .4429079
 cen_size100 |   .0367348  .0058584    6.27   0.000    .0252526    .048217
       _cons |   -.236393  .0261528   -9.04   0.000   -.2876515  -.1851346
-------------------------------------------------------------------------


-------------------------------------------------------------------------
  Random-effects Parameters  |   Estimate  Std. Err.   [95% Conf. Interval]
-----------------------------+-------------------------------------------
schoolid: Unstructured       |
                 sd(female)  |   .1436581  .0280289    .0980063   .2105749
                 sd(_cons)   |   .4440374  .0216798    .4035156   .4886284
         corr(female,_cons)  |  -.3851178  .1293113   -.6066708  -.1080499
-----------------------------+-------------------------------------------
               sd(Residual)  |   .8328283  .0051746    .8227477   .8430323
-------------------------------------------------------------------------

LR test vs. linear regression:     chi2(3) =  1777.38   Prob > chi2 = 0.0000
```

RESIDUAL DIAGNOSTICS

In Chapter 2, we conducted some checks of normality around our Level 1 and Level 2 residuals. In a random coefficient model, we have to inspect the distribution of the random coefficients as well. As for the residuals, the assumption is that they are normally distributed. Let us use our model with two random coefficients from Table 3.6:

```
mixed z_read cen_escs female || schoolid: cen_escs ///
   female, cov(un)

predict resid, residuals

predict u1 u2 u0, reffects
```

After the **mixed** command we use the post-estimation command, **predict**, to generate new variables for the Level 1 residuals (*resid*) and then for the first random effect for family capital (*u1*), for the second random effect for female (*u2*), and finally for the random intercepts or Level 2 residuals (*u0*).

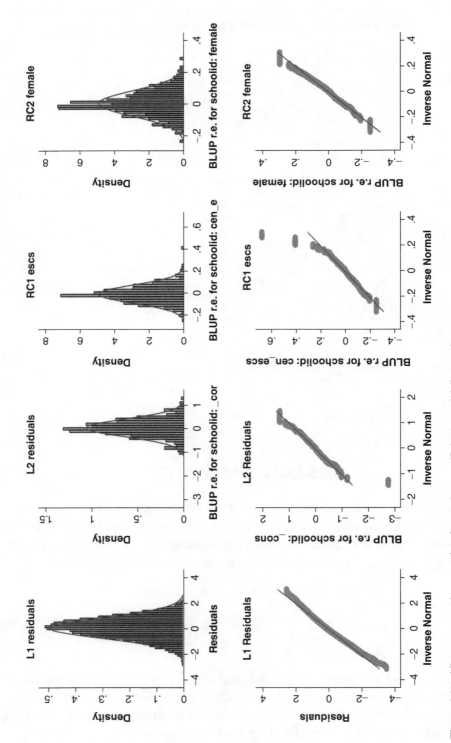

Figure 3.11 Histograms and qnorm plots for random coefficients model diagnostics

We show each of these four new variables as histograms and qnorm plots in Figure 3.11. The plots illustrate that the Level 1 residuals are normal. The Level 2 (intercept) residuals and RC1 (family capital), however, have some definite outliers. There is no agreed treatment of outliers, but we should investigate this further.

It is easy to identify the schools that are the outliers by simply listing the *schoolid* values for those schools that appear to have slopes of greater that 0.2 or intercepts that are less than –1.5:

```
list schoolid if u0 < -1.5 & pickoneone==1
list schoolid if u1 > 0.3 & pickone==1
```

```
. list schoolid if u0 < -1.5 & pickone==1

        +----------+
        | schoolid |
        |----------|
5144.   |      132 |
        +----------+

. list schoolid if u1 > 0.3 & pickone==1

        +----------+
        | schoolid |
        |----------|
4182.   |      107 |
5144.   |      132 |
        +----------+
```

School 132 comes up as an outlier in terms of both its intercept and slope. There are only six students in this school. If this were a real-life analysis, you may want to consider excluding this school. We can run the random coefficient model from Table 3.6 without school 132 and examine the outputs in Table 3.9.

The results for all schools in the sample are on the left and the results excluding school 132 are on the right. The fixed part of the estimation is rather similar. In the random parameter panel, however, we can see some noticeable differences: the values of the variances for the random coefficient for family capital and intercept are somewhat smaller when school 132 is excluded. The correlation between the random coefficient for family capital and intercepts also falls noticeably from –0.387 to –0.241, suggesting that the outlier was highly influential in inflating the correlation. You can also see the differences in rho between the two estimations; in the model with the outlier school 25% of the variation in standardized reading

Table 3.9 Comparison of results with and without school no. 132

	All schools	Exc. no. 132
cen_escs	0.354	0.354
female	0.404	0.404
_cons	−0.216	−0.211
var (cen_escs)	0.017	0.013
var (female)	0.022	0.023
var (_cons)	0.186	0.161
corr (cen_escs, female)	−0.276	−0.285
corr (cen_escs, _cons)	−0.387	−0.241
corr (female, _cons)	−0.479	−0.535
var (Residual)	0.672	0.672
N	13,644	13,638
rho	0.251	0.226
Log likelihood	−17,142.9	−17097.5

scores can be attributed to differences between schools, while in the model excluding the outlier this falls to 22.6%. These diagnostics point to the quite severe influence of a few cases – six students in one school!

FIRST STEPS IN MODEL-BUILDING

You may have gathered by now that we advocate a step-by-step approach to building your models where you test if each step is necessary with a view to keeping the model as parsimonious as possible. Coming from a general approach to data analysis of univariate then bivariate then multivariate, we advocate careful inspection and testing of your data at the univariate and bivariate stages before moving on to your multivariate models. The bivariate stage would include investigating null models with your dependent variable to determine if your grouping variable warrants multilevel modeling and then examining each independent variable within a bivariate multilevel model to determine if there is evidence to support modeling it as a random intercept or a random coefficient (which includes random intercepts). You may find from this process that some independent variables can be modeled with a single 'fixed' coefficient quite adequately. What we are saying here is that unless you go through this process, you won't have the foundation of information you need, once your model becomes more complex, to help with the interpretation.

We have used the **lrtest** command to assess if a model fits the data better than a previous model. This command requires that your sample sizes for both models are the same and that the first model is nested within the second. From your exploration of the data you should be able to arrive at a final estimation sample, which you can then use for all of your model testing.

In our examples, we have investigated the effects of parental occupational status, gender, and school size of reading scores with school as the nesting variable. As a set of analyses we would go about this in this way:

1. Null model
2. Add random intercept model Level 1 variables, one at a time and compare with null model
3. Add random coefficient model Level 1 variables, one at a time and compare with their random intercept model
4. Build the multivariate model:
   (a) Random intercept model, all Level 1 variables
   (b) Random coefficient model, all Level 1 variables (some/all as random coefficients) and compare with 4(a)
   (c) Random coefficient model, all Level 1 and Level 2 variables and compare with 4(b).

In Stata commands this could be done in this way:

```
* Step 1
mixed z_read || schoolid:, nolog
* see Table 2.9
est store null

LR test vs. linear regression: chibar2(01)  =  2675.02 Prob >= chibar2 = 0.0000


* Step 2(a)
mixed z_read cen_pos || schoolid:, nolog
* see Table 2.11
est store ri_pos
lrtest ri_pos null

Likelihood-ratio test                     LR chi2(1)        =      731.55
(Assumption: null nested in ri_pos)       Prob > chi2       =      0.0000


* Step 2(b)
mixed z_read female || schoolid:, nolog
est store ri_female
lrtest ri_female null

. lrtest ri_female null

Likelihood-ratio test                     LR chi2(1)        =      605.13
(Assumption: null nested in ri_female)    Prob > chi2       =      0.0000
```

```
* Step 3(a)
mixed z_read cen_pos || schoolid: cen_pos, cov(un) nolog
* see Table 3.1
est store rc_pos
lrtest rc_pos ri_pos
```

```
Likelihood-ratio test                        LR chi2(2)    =         1.24
(Assumption: ri_pos nested in rc_pos)        Prob > chi2   =       0.5391
```

```
* Step 3(b)
mixed z_read female || schoolid: female, cov(un) nolog
* see Table 3.5
est store rc_female
lrtest rc_female ri_female
```

```
Likelihood-ratio test                        LR chi2(2)    =        15.06
(Assumption: ri_female nested in rc_female)  Prob > chi2   =       0.0005
```

```
* Step 4(a)
mixed z_read cen_pos female || schoolid:, nolog
est store mod_a
```

```
* Step 4(b)
mixed z_read cen_pos female || schoolid: female, cov(un) nolog
* see Table 3.7
est store mod_b
lrtest mod_b mod_a
```

```
Likelihood-ratio test                        LR chi2(2)    =        13.78
(Assumption: mod_a nested in mod_b)          Prob > chi2   =       0.0010
```

```
* Step 4(c)
mixed z_read cen_pos female cen_size100 || ///
    schoolid: female, cov(un) nolog
* see Table 3.8
est store mod_c
lrtest mod_c mod_b
```

```
Likelihood-ratio test                        LR chi2(1)    =        37.12
(Assumption: mod_b nested in mod_c)          Prob > chi2   =       0.0000
```

As you can see from the notes in the commands, most of the tables have already been shown early in this chapter or Chapter 2. We have included the **lrtest** results here to help the discussion.

At Step 1 we see that the random intercept null model significantly improves the fit of the model over the single-level regression model.

In both stages of Step 2, the addition of the Level 1 variables – parental occupational status and gender – significantly improves the model fit over the null model.

In Step 3(a) when parental occupational status is modeled as a random coefficient, there is no significant improvement in model fit compared to when it is modeled as a random intercept only. Therefore we would want to model parental occupational status as a 'fixed' effect. Parental occupational status has a significant effect on reading scores, which is shown in the coefficient in the fixed part of the output, but that coefficient does not vary significantly across schools.

Also in Step 3(b) we see that modeling gender as a random effect significantly improves the model fit compared to when it is modeled as a random intercept only. Therefore we would want to keep gender as a random effect as the effect of gender varies significantly across schools.

In Step 4(a) we model the two Level 1 independent variables in a random intercept model. When entered together, both still have a significant 'fixed' effect on reading scores. We save the estimates to use for comparisons with later models.

In Step 4(b) we model the two Level 1 independent variables in a random coefficient model where gender is in both the fixed and random part of the command and parental occupational status is only in the fixed part. We see that by doing this the model fit is significantly improved compared to the model in Step 4(a).

In Step 4(c) we enter the Level 2 variable in the fixed part of the command. We see that this also significantly improves the model fit.

By taking this step-by-step approach, we have ended up with the least complex model and developed our understanding of how best to model the variables along the way. You could go on to further steps where you investigate interactions between the variables depending on how they relate to your research questions. Our takeaway point here is that it is best to build up slowly to more complex models in order to understand what is happening with each of your independent variables, rather than just dive straight in with a complex model that is difficult to unpick.

SOME TASTERS OF FURTHER EXTENSIONS TO OUR BASIC MODELS

We briefly introduce some other variations/extensions to the basic two-level linear models we have covered so far. We start with a look at three-level models using schools at Level 2 and regions at Level 3. Then we move on to dichotomous dependent variables, a quick note on cross-classified models, and finish with weighting.

Three-level models

In this book we have concentrated on two-level models so that new users can more easily visualize and understand what is happening, why they might choose these models, and how to interpret the output. We believe that spending time on the basics or fundamentals sets a good foundation for expanding your knowledge through further reading and practice. So, as a taster for how to start out with three-level models we continue with our previous example where we are investigating the effect of parental occupational status, gender and school size on reading scores.

In Step 4(c) above we had a two-level model with two Level 1 (student characteristics) variables – parental occupational status and gender – and one Level 2 (school characteristics) variable. With a two-level model we could specify the Level 1 variables in the random and/or fixed parts of the command, but we could only specify the Level 2 variable as a 'fixed' effect as there was no variation within schools, only between.

We now expand this model into three levels where we nest schools within regions, which allows us to model the Level 2 variable as a random coefficient across regions. We still use the **mixed** command and in the random parts, after the fences, we specify the levels. The highest level (*region*) is added first, followed by the school level (*schoolid*). So, the null model looks like this:

```
mixed z_read || region: || schoolid: , nolog
```

We run this and test the model fit against the two-level model (Step 1 above).

In the output shown in Table 3.10, there is a table with the descriptives of the grouping variables. As we have seen from previous output, there are eight regions, within those eight regions are 356 schools, and within those schools are 13,646 students. The average observations per group are Level 1 units – students – so there is an average of 1,706 students per region and an average of 38 students per school. If you wanted to see how many schools there were per region then you could use:

```
tabstat schoolid if pickone==1, by(region) stat(n)

region   |    N
---------+-------
    ACT  |   26
    NSW  |   80
    VIC  |   22
    QLD  |   57
     SA  |   43
     WA  |   34
    TAS  |   56
     NT  |   38
---------+-------
  Total  |  356
-----------------
```

Table 3.10 Stata output for three-level null model, region and school groups

```
Mixed-effects ML regression                    Number of obs    =     13646


-----------------------------------------------------------
               |  No. of      Observations per Group
Group Variable |  Groups   Minimum   Average   Maximum
---------------+-------------------------------------------
        region |       8       702    1705.8      3270
      schoolid |     356         3      38.3        57
-----------------------------------------------------------


                                          Wald chi2(0)      =        .
Log likelihood = -18010.269               Prob > chi2       =        .


-----------------------------------------------------------------------
      z_read |     Coef.   Std. Err.     z   P>|z|    [95% Conf. Interval]
-------------+---------------------------------------------------------
       _cons | -.0732865    .092708  -0.79   0.429   -.2549909   .1084179
-----------------------------------------------------------------------


-----------------------------------------------------------------------
  Random-effects Parameters |  Estimate   Std. Err.   [95% Conf.  Interval]
----------------------------+------------------------------------------
region: Identity            |
                var(_cons)  |  .0620496   .0365245   .0195746    .1966914
----------------------------+------------------------------------------
schoolid: Identity          |
                var(_cons)  |  .2323436   .0201353   .1960487    .2753579
----------------------------+------------------------------------------
             var(Residual)  |  .7676147   .0094264   .7493598    .7863143
-----------------------------------------------------------------------
LR test vs. linear regression:    chi2(2) =  2704.13    Prob > chi2 = 0.0000

Note: LR test is conservative and provided only for reference.

. est store null3

. lrtest null3 null

Likelihood-ratio test                         LR chi2(1)    =      29.11
(Assumption: null nested in null3)            Prob > chi2 =      0.0000
```

The rest of the output is very similar to the two-level null model except that in the random effects parameters section there are two variances for random intercepts – one at Level 2 for school (*schoolid*) and one at Level 3 for region. The model fit is significantly improved over a single-level regression model and over the two-level null model.

Next, we specify a three-level random intercept model and compare the model fit with the three-level null model.

```
mixed z_read cen_pos female cen_size100 || ///
        region: || ///
        schoolid: , nolog nogroup
```

Table 3.11 Stata output for three-level random intercept model of reading scores regressed on parental occupational status, gender and school size

```
Mixed-effects ML regression                    Number of obs    =      13646
                                               Wald chi2(3)     =    1491.74
Log likelihood = -17308.817                    Prob > chi2      =     0.0000


------------------------------------------------------------------------------
     z_read |     Coef.   Std. Err.      z    P>|z|    [95% Conf. Interval]
------------+-----------------------------------------------------------------
    cen_pos |  .0134239   .0004791    28.02   0.000    .0124849    .0143629
     female |  .4085257   .0159766    25.57   0.000    .3772121    .4398393
cen_size100 |  .0350726   .0059731     5.87   0.000    .0233655    .0467797
      _cons | -.2609924   .0743225    -3.51   0.000   -.4066619   -.1153229
------------------------------------------------------------------------------


------------------------------------------------------------------------------
  Random-effects Parameters  |  Estimate   Std. Err.    [95% Conf. Interval]
-----------------------------+------------------------------------------------
region: Identity             |
                 var(_cons)  |  .0390276   .0233941    .0120542    .1263581
-----------------------------+------------------------------------------------
schoolid: Identity           |
                 var(_cons)  |  .1563351   .0143364    .1306167    .1871174
-----------------------------+------------------------------------------------
               var(Residual) |  .6975337   .0085726    .6809325    .7145396
------------------------------------------------------------------------------
LR test vs. linear regression:     chi2(2) = 1792.51   Prob > chi2 = 0.0000

Note: LR test is conservative and provided only for reference.

. est store ri3

. lrtest ri3 null3

Likelihood-ratio test                             LR chi2(3)   =    1402.90
(Assumption: null3 nested in ri3)                 Prob > chi2  =     0.0000
```

We see from the output in Table 3.11 that all three independent variables have a significant effect on reading scores and the model fit is significantly improved over the three-level null model.

Following on from what we already know about these independent variables, we move on to a random coefficient model where gender is specified as a random coefficient at Level 2 (school). We use the **stddev** option to produce standard deviations rather than variances and correlations instead of covariances so we can examine the correlations in the output. Then we compare this to the previous model.

```
mixed z_read cen_pos female cen_size100 || ///

         region: || ///

         schoolid: female, cov(un) stddev nolog nogroup
```

Table 3.12 Stata output for three-level random coefficient model specifying gender as a random coefficient at Level 2

```
Mixed-effects ML regression                    Number of obs     =      13646
                                               Wald chi2(3)      =     1343.14
Log likelihood = -17302.782                    Prob > chi2       =      0.0000


-------------------------------------------------------------------------
      z_read |     Coef.   Std. Err.     z    P>|z|    [95% Conf. Interval]
-------------+-----------------------------------------------------------
     cen_pos |   .0134097  .0004789   28.00   0.000    .0124711    .0143483
      female |   .4065644  .0180384   22.54   0.000    .3712098     .441919
 cen_size100 |   .0340731  .0059156    5.76   0.000    .0224788    .0456674
       _cons |  -.2626461  .0735643   -3.57   0.000   -.4068295   -.1184627
-------------------------------------------------------------------------


-------------------------------------------------------------------------
  Random-effects Parameters |  Estimate  Std. Err.    [95% Conf. Interval]
----------------------------+--------------------------------------------
region: Identity            |
              sd(_cons) |   .1942409  .0583297    .1078273    .3499072
----------------------------+--------------------------------------------
schoolid: Unstructured      |
             sd(female) |   .1443955   .027948    .0988102    .2110111
              sd(_cons) |   .4151287  .020839     .3762299    .4580493
      corr(female,_cons) |  -.3945607  .1280411   -.6134399   -.1193902
----------------------------+--------------------------------------------
           sd(Residual) |   .8327979  .0051736    .8227193        .843
-------------------------------------------------------------------------
LR test vs. linear regression:    chi2(4) =  1804.58   Prob > chi2 = 0.0000

Note: LR test is conservative and provided only for reference.

. est store rc3_1

. lrtest rc3_1 ri3

Likelihood-ratio test                          LR chi2(2)    =       12.07
(Assumption: ri3 nested in rc3_1)              Prob > chi2 =        0.0024
```

We see from the output in Table 3.12 that the *female* variable has a moderate negative correlation with the Level 2 random intercept, suggesting a fanning-in of the regression lines, which we have seen previously in the two-level models. Specifying gender as a random coefficient significantly improves the model fit compared to the random intercept model.

In the two-level model we were only able to specify the Level 2 variable (school size) – as a 'fixed' effect as it is a constant within each school – all students in a school have the same value for school size. When we say 'only able' we mean in a meaningful way. Of course, you can put this in the two-level model Stata command and it will produce output. But what would it mean? Within each school, school size is a constant so you cannot obtain a correlation or regression line for every individual school, so how would the coefficients meaningfully vary across schools? In this three-level model we can specify school size as a random coefficient at Level 3 (region) and compare it to the previous model. Note that because now we have two random effects at both school and region levels we need to specify the covariance structure – **cov(un)** – for each one in the **mixed** command.

```
mixed z_read cen_pos female cen_size100 || ///

    region: cen_size100, cov(un) || ///

    schoolid: female, cov(un) stddev nolog nogroup
```

Table 3.13 Stata output for three-level model specifying gender as a random coefficient at Level 2 and school size as a random coefficient at Level 3

```
Mixed-effects ML regression              Number of obs     =      13646
                                         Wald chi2(3)      =    1310.44
Log likelihood = -17290.158              Prob > chi2       =     0.0000
```

z_read	Coef.	Std. Err.	z	P>|z|	[95% Conf.	Interval]
cen_pos	.0134905	.0004783	28.20	0.000	.012553	.0144279
female	.4049526	.0180087	22.49	0.000	.3696562	.4402491
cen_size100	.0529234	.0204192	2.59	0.010	.0129026	.0929443
_cons	-.2071695	.0339807	-6.10	0.000	-.2737705	-.1405686

Random-effects Parameters	Estimate	Std. Err.	[95% Conf.	Interval]
region: Unstructured				
sd(cen_~100)	.0549396	.0166472	.0303364	.0994963
sd(_cons)	.0593754	.0385402	.0166378	.211893
corr(cen_~100,_cons)	-.3136969	.5891581	-.9224779	.7425423
schoolid: Unstructured				
sd(female)	.1442984	.0279204	.0987556	.210844
sd(_cons)	.4068213	.0202575	.3689932	.4485275
corr(female,_cons)	-.4812035	.1185626	-.6788354	-.218564
sd(Residual)	.8326527	.005171	.8225792	.8428495

```
LR test vs. linear regression:   chi2(6) =   1829.83   Prob > chi2 = 0.0000

Note: LR test is conservative and provided only for reference.

. est store rc3_2

. lrtest rc3_2 rc3_1

Likelihood-ratio test                        LR chi2(2)   =      25.25
(Assumption: rc3_1 nested in rc3_2)          Prob > chi2  =     0.0000
```

We see that including the random coefficient for school size across regions significantly improves the model fit (Table 3.13). The moderate negative correlation

(–0.314) between the random intercept for *cen_size100* and the Level 3 random intercept suggests a fanning-in of the regression lines, which in turn suggests more variance in reading scores among smaller schools than larger schools. However, with only eight regions this is somewhat suggestive.

Dichotomous dependent variables

Another extension of the models we have presented in this book is to model a dichotomous or binary dependent variable in an analogous way to the single-level logit or logistic regression. The multilevel command in Stata is **melogit** and it follows the same structure as **mixed** with a fixed and random part to the command. For example, if we created a dichotomous dependent variable that indicated the highest 10% on reading scores – *hiread* – and modeled parental occupational status and gender using school groups, we would use for a random intercept model:

```
melogit hiread cen_pos female || schoolid:
```

The output is shown in Table 3.14. A model that specifies the variable *female* as a random effect would be:

```
melogit hiread cen_pos female || schoolid: female, or
```

The option **or** can be used to produce odds ratios in the output: exp(b). We should note that our terminology regarding random intercept and random coefficient models gets rather strained with dichotomous dependent variables, but we stick with it for this brief section.

Stata takes considerably longer to estimate these models. The random intercept models are done reasonably quickly, but when random coefficients are introduced the time increases quite dramatically. For example, the simple random coefficient model above took over 50 iterations to converge on the estimation. You may have noticed that we did not specify the **cov(un)** option in this example. This option dramatically increases the number of iterations needed to converge. See for yourself and run the code in the do-files and data that accompany this book. It was only a matter of minutes in time, but you can see how complex models would take a good amount of time to converge at estimation.

From Table 3.14 you should quickly see that there is no reason to model gender as a random coefficient and it may be better to use a random intercept model in this case. The output is very similar to that produced by the **mixed** command, with two main sections – one for the fixed effects and one for the random effects. The fixed effects section is interpreted as in ordinary logistic regression. The random effects section shows the variance of the random coefficient for the variable

Table 3.14   Stata output for dichotomous dependent variable specifying gender as a random coefficient using `melogit` command

```
Mixed-effects logistic regression          Number of obs      =        13644
Group variable: schoolid                   Number of groups   =          356

                                           Obs per group: min =            3
                                                          avg =         38.3
                                                          max =           57

Integration method: mvaghermite            Integration points =            7

                                           Wald chi2(2)       =       369.22
Log likelihood = -3856.6317                Prob > chi2        =       0.0000
-------------------------------------------------------------------------------
      hiread | Odds Ratio  Std. Err.      z    P>|z|     [95% Conf. Interval]
-------------+-----------------------------------------------------------------
     cen_pos |   1.035741   .0023136    15.72   0.000     1.031216    1.040285
      female |   2.375917   .1784381    11.52   0.000     2.050708    2.752701
       _cons |   .0391616   .0033841   -37.50   0.000     .0330603    .046389
-------------+-----------------------------------------------------------------
schoolid     |
 var(female) |   4.21e-34   1.53e-18                        .           .
 var(_cons)  |   .9479076   .122949                       .7351236    1.222283
-------------------------------------------------------------------------------
LR test vs. logistic regression: chibar2(01) = 471.62 Prob >= chibar2 = 0.0000
```

female, var(female), and for the intercept, var(_cons). The variance for *female* is zero, which suggests removing this random effect from the model. The other main Level 1 variables that we have used in our examples – parental occupational status and family capital – also do not return significant random effects in these dichotomous dependent variable models. There is no covariance between the random effect of *female* and the random intercept as the covariance is specified as zero because we left out the **cov(un)** option.

Cross-classified models

Cross-classified models are used when the nesting of the data is more complicated than what we have considered so far. In our models, we have assumed that lower-level units belong to a single higher-level unit. Students (lower level) attend a school (higher level), but if a researcher wants, for example, to explore the topic of neighbourhoods and schools, this is where cross-classified models come into play. Not all students in a neighbourhood attend the same school, so if you have information on a student's neighbourhood, all students in a school won't necessarily be from the same neighbourhood, nor will all the students in a single neighbourhood attend the same school. Figure 3.12 provides an illustration of how nesting structures can be cross-classified. You can easily identify if your data require analysis by cross-classified models if the nesting levels you are interested in, such as neighbourhood and school, are not 'cleanly' divided up so that multiple category memberships are possible. A simple cross-tabulation between neighbourhood and schools would reveal this in your initial data exploration. The data explored in our analysis data set do not have such characteristics, however. Readers interested in further exploration of cross-classified models are encouraged to read Leckie (2012) for an accessible discussion of how (and why) to undertake such analyses, as well as Leckie (2013) for a worked example using Stata.

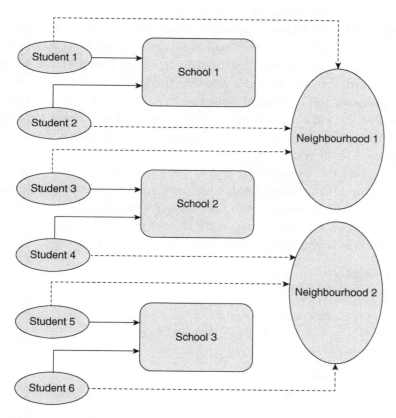

Figure 3.12 An example of a cross-classified model

Weighting

Unless data are drawn from a random sample of the population, it is often necessary to use weighting in your analysis. Until version 12 of Stata, weighting was not possible in the **xtmixed** command, which limited its utility with complex data structures. In secondary data sources, such as PISA, weights are provided in the data set and are fully explained in the supporting documentation. This is the case with many large and reputable data sources. Having said that, it is up to you to figure out the weighting issues in your data. There is not a one-size-fits-all solution and often it is necessary to consult on the proper use.

In the Australian PISA data, there are student weights and school weights. These are necessary because the schools and students were not selected as part of a simple random sample. The schools and students had different probabilities of selection as well as varying non-participation rates for different school groups or student characteristics. Some types of schools were also oversampled to help answer national policy questions (OECD, 2009).

Options within the **mixed** command allow you to weight at all the levels of your model. Unlike other commands in Stata, weights in **mixed** need to be 'scaled'. In layman's terms, you need to tell Stata how to treat the Level 2 weights in relation to your Level 1 units. In the following example, we use the option **size** with the command **pwscale** to specify that Level 1 weights are to be rescaled so that they add up to their Level 2 group. There are two other options and you should check the Stata help files to help you decide what is most appropriate for your data.

Below we run the unweighted and weighted estimations for our previous Model 3 and present the coefficients together using the **esttab** command – you will need to install this user-written package. We needed to convert how Stata stores estimates back to the values in the output within the **esttab** command. We just note this here and return to producing tables in Chapter 4.

```
mixed z_read female cen_pos cen_size || schoolid:
estimates store unweighted

mixed z_read female cen_pos cen_size || ///
     schoolid: ,pweight(w_fschwt) pwscale(size)
estimates store weighted

esttab unweighted weighted, se b(4) se(4) ///
     transform(ln*: exp(2*@) 2*exp(2*@)) ///
     eqlabels("" "var(Constant)" "var(Residual)") ///
     mtitles("Unweighted" "Weighted")
```

Weighting has changed the values of the coefficients and the constants (see Table 3.15), indicating that the weights are performing an important job in our data! Although the weighted and unweighted results are not tremendously different, it is important to use weights where they are provided in order to adjust for the sampling structures within your data.

There is one thing to note on how we have used centred variables and weights. The variables were centred on the unweighted sample mean, which, more likely than not, is different from the weighted sample mean. Therefore, to interpret the

Table 3.15 Comparison of unweighted and weighted estimates

```
-------------------------------------------
                        (1)             (2)
                  Unweighted        Weighted
-------------------------------------------

female             0.4096***       0.4511***
                  (0.0160)        (0.0274)

cen_pos            0.0134***       0.0130***
                  (0.0005)        (0.0006)

cen_size           0.0004***       0.0002*
                  (0.0001)        (0.0001)

_cons             -0.2360***      -0.2253***
                  (0.0251)        (0.0286)
-------------------------------------------
var(Consta~"
_cons              0.1801***       0.1550***
                  (0.0162)        (0.0319)
-------------------------------------------
var(Residu~)
_cons              0.6976***       0.6930***
                  (0.0086)        (0.0162)
-------------------------------------------
N                     13646           13646
-------------------------------------------
Standard errors in parentheses
* p<0.05, ** p<0.01, *** p<0.001
```

intercept in the weighted model you have to use the actual figures used to centre the variables rather than saying 'the average'. Another approach to centring variables is to pick a value which makes sense theoretically and/or practically and centre around that. In this way, the interpretations are the same for the unweighted and weighted model.

WHERE TO NEXT?

As we said at the beginning, this book is aimed at those starting out with no previous experience of multilevel models. Once you have grappled with the basic concepts of these models and understood the rationale for using them, a whole world of modeling approaches awaits. And things can get very complicated very quickly! As with single-level models, there are ways to estimate models with various types of dependent variables. We briefly mentioned the **melogit** command for dichotomous dependent variables, but there is also the **meprobit** command. For count dependent variables there are **mepoisson** and **menbreg** commands. After these, you may want to progress onto looking at the `gllamm` command, which you need to install and is thoroughly covered in Rabe-Hesketh and Skrondal (2012).

———————— CHAPTER 3 TAKEAWAY POINTS ————————

- Random coefficient models allow your Level 2 groups to have different slopes.
- Random coefficient models usually assume (unless otherwise specified) that the intercepts are allowed to vary as well. Thus, the Level 2 groups will have varying slopes *and* intercepts.
- Your theoretical framework should rationalize your choice for using random coefficients. They should not just be added to be 'fancier' because they complicate the interpretation of your models significantly.
- Likelihood ratio tests should be used in order to determine if your model fit has been improved by adding a random coefficient.
- You must explore the correlation between your random intercepts and coefficients in order to make sense of the relationships being demonstrated in your data.

FOUR

Communicating Results to a Wider Audience

CHAPTER CONTENTS

In this final chapter, we focus on how to present your multilevel model results. We have taken you through various steps of exploring the relationships in your data. We have shown you how to interpret the various outputs produced by Stata and also shown you, with the assistance of graphics, what the models are telling you.

As a researcher, one of your tasks is to communicate your results to a wider audience – an audience that, in many cases, may not have an understanding of multilevel modeling, but may also not have a mastery of basic statistics either. Beyond understanding your output and what your results mean, to write an effective report or paper you have to be able to summarize your results in a meaningful way. What a 'meaningful way' is will depend on your audience, of course. A non-technical audience will need a different approach than, say, an anonymous reviewer of a peer-reviewed journal that publishes a lot of quantitative articles.

CREATING JOURNAL-FORMATTED TABLES

We are firm believers that the raw output produced by statistical software packages should never be copied and pasted into presentations, reports, dissertations, theses, or papers. Such output typically gives *much more* information than any reader requires. It is your job as the communicator of results to summarize your findings for your audience – not your audience's job to sift through masses of information to figure out what is relevant.

But what is relevant? New researchers often don't know what to report because software programs report so much information, as do books on how to do multilevel modeling (including this one). There is often an assumption that *more* is better. Much of the information we have provided in this book by way of examples and graphical images is to help you understand what it is you are doing when you create multilevel models – they are not included as steps you should follow in producing a final report. They are included to show you what the data are doing, in the hope that such information will help you understand the bigger picture behind multilevel modeling techniques. For example, it is not necessary to produce charts of residuals, caterpillar plots and the like for a publication in a scholarly journal. These plots will help you get familiar with your data, but no one needs to see them apart from the researchers working on the project.

THE FIXED PART OF THE MODEL

As you may have noticed from the previous chapters of this book, much of the information reported by Stata is somewhat repetitive. For example, in every regression the results report the coefficient, the statistic of significance associated with the coefficient (e.g. the t statistic), the p value, and the confidence interval. With regard to the coefficient and its significance, the latter three pieces of information overlap in their usefulness. It is not necessary to report the t statistic, the p value and the confidence interval! Often, researchers take away all these pieces of information and simply report asterisks and similar symbols (along with a legend at the foot of the table indicating their meaning) to demarcate statistical significance.

For example, the following fixed part of a regression is from Table 2.14:

```
------------------------------------------------------------------
 z_read |    Coef.    Std. Err.     z    P>|z|  [95% Conf. Interval]
---------+--------------------------------------------------------
 cen_pos | .0133781    .0004791   27.92  0.000    .012439   .0143172
  female | .4095749    .0159999   25.60  0.000   .3782156   .4409341
cen_size | .0003781    .000059     6.40  0.000   .0002624   .0004938
   _cons |-.2360258    .0250803   -9.41  0.000  -.2851823  -.1868692
------------------------------------------------------------------
```

To report these fixed parts in a table, just like any 'normal' regression, you could reduce them to:

Independent variables	Coefficient
Parental occupation status (centred)	0.0134***
Female	0.4096***
School size (centred)	0.0004***
Constant	−0.2360***

*$p < 0.05$, **$p < 0.01$, ***$p < 0.001$

It should be noted that this preference for what to report is rather discipline, and journal, specific. The above example is a style that is widely accepted in most social sciences, but your discipline's practice may be different, such as preference for standard errors to be reported instead of (or in addition to) the asterisks. Also note that in multilevel models, the 'constant' is often reported first and labelled the 'intercept':

Independent variables	Coefficient
Intercept	−0.2360***
Parental occupation status (centred)	0.0134***
Female	0.4096***
School size (centred)	0.0004***

*$p < 0.05$, **$p < 0.01$, ***$p < 0.001$

However, as you know, there are more parts to a multilevel model, namely the random components. You will also have to report these in your table.

What you discuss and how you discuss it, however, will depend entirely upon your hypotheses. Often – very often indeed – researchers use multilevel models simply to control for the clustering within their data. They use multilevel models because they don't want to violate the assumptions of OLS. This, incidentally, is how we (the authors of this book) have used these models in our own research. We tend to report random components in the tables, but we don't talk much about the different random components because *our research question is not concerned with it.* It is absolutely fine to focus on the fixed part of your model as you would in a normal regression and simply report the random components without giving them a whole lot of attention if simply accounting for the clustering in your data is the only reason you were using multilevel models in the first place.

It is also quite common to present models in stages. In this approach, various estimations are produced in separate columns, with a final estimation being presented in the last column. One reason for using this technique is to test how different groups of variables contribute to the model, which can be more readily observed when the variables are added in batches. For instance, if you wanted to see if a variable of interest had a mediating effect on your model – and by 'mediating' we mean that its influence was partially explained by changes to other variables in your model that you observed upon adding the mediator – you could add it in a separate stage.

A common approach for reporting the results from multilevel models is:

1. Present the null model.
2. Present the model with only the main hypothesized variable of interest. In other words, if your hypotheses are focused on cultural capital, for example, show the results for your variable of cultural capital.
3. Add literature-recommended control variables.
4. Sometimes variables measuring Level 2 characteristics are added at a separate stage (if not added in Step 3).
5. Sometimes interaction effects – if theoretically relevant – are added in a separate step.

Wherever you stop adding variables is your final model. Sometimes researchers also present models in other various permutations, but there should be a valid reason for doing so that logically stems from their discussion and hypotheses.

Bell (2002) and Pike and Rocconi (2012) offered some very straightforward suggestions on how best to present multilevel results. Bell's suggestions were primarily intended for the presentation of multilevel models, but much of their logic extends to many types of quantitative model presentation.

1. Several models within the same table should be clearly labelled. Rather than simply labelling models 'Model 1', 'Model 2', etc. – or with roman numerals – it is much more informative to the reader to indicate what the models include. For example, for a table including a null model and then adding Level 1 variables, then Level 2 variables, it is rather easy to indicate these additions in the column titles.
2. When using categorical variables, it is important to indicate your base category somewhere. Don't leave it to the reader to figure it out. Be explicit.
3. Don't 'overly round' the figures in your tables. Leave at least three decimal places.
4. If a particular variable is the focus of your analysis – that is, you are interested in how a particular independent variable is impacted by the inclusion of additional variables – it should be as easy as possible to see the changes in coefficients.
5. Random variation is understood more easily as standard deviations than as variances.
6. The significance values should always be explained. In other words, what do your stars mean?

7. Where possible, it is good to display technical results graphically. A well-designed graph should be able to tell a major part of your statistical story with a single image.

We would also like to add an eighth suggestion. If you are using logistic regression, presenting odds ratios is *almost always* a better idea than presenting logistic regression coefficients. Odds ratios are far easier to understand.

THE IMPORTANCE OF THE NULL MODEL

While researching this book, we surveyed the recent literature in 'top' journals in the social sciences to examine the current practice in reporting multilevel models. We noticed that the presentation of the null model as the first step appears to be falling out of fashion. And statistics have their fashions – make no mistake. Up until about 15 years ago, it would have been unthinkable to present regression results without also presenting a correlation matrix to demonstrate that the independent variables weren't overly correlated with one another. Doing so today would be considered redundant and perhaps even laughable.

This is possibly the direction that the presentation of the null model as the 'first step' is taking, but we still believe that it is a rather useful piece of information to have. Sometimes the null model is used to rationalize the choice of using multilevel models in the first place. Remember that if your rho is small (some say less than 0.10), you are not really adding anything to your analysis by doing multilevel modeling because there is not much variance in the levels of your data. In such cases, using single-level OLS would be just fine. First and foremost, the null model tells you the variance in the model that is explained by your Level 2 groups. So when we do research on schools, for example, we can know, before undertaking any significant model-building, how much of the variance in our dependent variable can simply be attributed to differences between schools. As you add variables (or groups of variables), you can see how this changes. In terms of how many researchers present and discuss the null model compared to those who do not, it varies considerably. It is our opinion this information should be revealed to readers if not in a step in model-building explicitly, then at the very least as a footnote.

CENTRING VARIABLES

In the previous chapters, we've focused on centring variables to make the intercepts easier to understand. Rather, centring makes the intercepts *possible* to understand in many cases where variables don't have meaningful zero values. However, in practice, researchers very rarely centre their variables. When you publish results in a journal, you may find it unnecessary to centre your variables. In fact, you might find it prob-lematic to do so (this is the authors' experience) unless you specifically want to discuss the intercepts in your results. Reviewers and editors may ask you not to centre vari-ables! Discussing the intercepts may be the convention in some disciplines, so it is probably best if you ascertain from your own search of the journals in your area what is current best practice. We do not centre the variables in the examples in this chapter.

STATA COMMANDS TO MAKE TABLE-MAKING EASIER

In Stata, there are commands to convert your regression output into journal-for-matted tables, which saves you the time of manually copying and pasting various pieces of information. In Stata, journal (almost)-ready tables can be produced quite easily through the `esttab` command. This command is part of a larger suite of commands in the estout.ado package that you may need to download as it does not come pre-loaded in Stata. In order to install the estout package, all you need to do is type in the command window:

```
ssc install estout
```

In the commands that follow below where `esttab` is used, the table appears onscreen within the Stata output window. You can also use the `using` option to save the table in a variety of file formats (.doc, .rtf, .html, etc.) where it is easier to edit them.

In the following worked example, we have a model that is presented in three stages: first, the null model; second, the model with *female* as the only inde-pendent variable; and finally, the model with *female* and three other variables. A researcher setting up his or her models this way could rationalize this manner of model-building by arguing first that the data under consideration showed significant variance at the school level (null model). To focus on females as the single variable of interest in the second model would be due to having hypoth-eses around the topic of gender and standardized reading score. The addition of the other variables in the third model would be informed by theory and/or previous research.

Let's begin with an example. Below we run a null model with reading scores as our dependent variable and individual schools as our levels. In the second model we add *female* as a sole independent variable, and then in the third model three additional independent variables – parental occupational status (*pos*), percentage of female students (*pcgirls*), and school size in units of 100 students (*schsize100*). One problem in previous examples is the very small coefficient for the variable for school size. It is often reported as 0.000 and statistically significant. This means that each additional student in a school improves reading scores by a very, very small amount – so small that it is 0 to the third decimal place. In order to make this value more meaningful, we have changed the scale of this variable by dividing the original value by 100 so that that the resulting coefficient is interpreted as the increase in reading scores for every 100 additional students.

```
mixed z_read || schoolid:
estat icc
estadd scalar icc = r(icc2)
estimates store null

mixed z_read female || schoolid:
estat icc
estadd scalar icc = r(icc2)
estimates store model2

mixed z_read female pos pcgirls schsize100 || schoolid:
estat icc
estadd scalar icc = r(icc2)
estimates store model3
```

You will notice that after **estat icc** we have included the command **estadd scalar icc = r(icc2)**. This is because in Stata 13 using **estat icc** produces the ICC as an "r-class" returned result. It is in this step that we are 'storing' the ICC in a manner that we can easily recall for producing tables. In Stata, commands that produce statistical results are either "r-class" or "e-class". The important distinction is that if they present estimations, they are "e-class". Only the e-class type of 'returned results' are *easily* added to an **esttab** command by way of the **scalar** option. An important distinction between r-class and e-class returned results is that they are stored in different places. Without digressing too much, this simply means that how you refer to these results in your commands will differ. If you use **return list** after **estat icc**, you can see the kinds of results it stores. Similarly, if you use **ereturn list** after an estimation command ('e' for estimation), you can see the types of results it stores. Quite simply, adding e-class results to an **estout/esttab** command is just easier because it is stored more conveniently for this type of task; r-class results aren't impossible to present – they just require a bit more work. Thus, this additional command just tells Stata to add the ICC that was calculated in the **estat** command and store it in memory as an r-class scalar called *icc2*. If you are using earlier versions of Stata you will find that **xtmixed** and **xtmrho** store the ICC as an e-class scalar.

```
esttab null model2 model3, se aic bic scalars(icc ll df_m) ///
        transform(ln*: exp(@)^2 exp(@)^2) ///
        eqlabels("" "var(Constant)" "var(Residual)" )
```

Now in the **esttab** command above we have specified the options for standard errors (**se**), AIC (**aic**), BIC (**bic**) and the 'estimation scalars' held in the memory of Stata that correspond to log likelihood (**ll**), the ICC and degrees of freedom (**df_m**).

The Level 1 variance (**var_u1**) and Level 2 variance (**var_e**) are also stored as scalars, but also stored in the matrix list that is accessed by command **estout**. They were not specifically requested in this command because they come up automatically as estimation parameters (although as the logarithms of standard deviations). However, you can access them in other commands using these scalar references. You can see the list of estimation scalars stored in temporary memory by typing **ereturn list** after any multivariate estimation. Any of these scalars can easily be added to the tables below by including them in the list of scalars in the **esttab** command.

In the **esttab** command you will also see the option **transform(ln*: exp(@)^2 exp(@)^2)**. Variance parameters of the **mixed** command are stored in Stata's memory as the logarithm of the standard deviation. In order to display them properly, we must tell Stata to transform them. In the **transform** option, the stored values are exponentiated and squared to be converted to the variances that are displayed in our **mixed** output. If you run the command *without* the **transform** option, you will notice that your variances will not match what is on your screen. Thus, the **transform** option isn't really optional for your output to make sense! The abbreviation **ln*** is used to refer to these variances as they are held in Stata's matrix memory as various coefficients that begin with 'ln'.

In the Stata matrix, *lns1_1_1_con* is the natural log of the standard deviation of the Level 1 errors, and *lnsig_e_cons* is the natural log of the standard deviation of the Level 2 random effect. These are the only two items in the matrix that start with 'ln', therefore denoting **ln*** tells Stata to make the exponentiation transformation with everything in the matrix containing this ln prefix. If you use **matlist e(b)** after a **mixed** command you will see all the coefficients held in a matrix in Stata's memory for the immediately preceding estimation.

Below are the results of our three models.

As you can see, this technique puts all your coefficients into a table with three separate columns, which is rather handy. Of course, you wouldn't simply copy and paste this into your document unedited. You need to label the columns for your different models and add some consistency to the number of decimals being presented. The amount of information presented in this table might still be a bit excessive. You may not need AIC, BIC and log likelihood. All three tell you if the model has been improved from the previous step. Pick the one most common in your field. Log likelihood is our favourite. If you look at multilevel results that have been published in different academic journals, you will see that there is no general 'industry standard' for how to present results.

This is obviously not quite a perfect table, but it is at least easily editable. There are many options you can play around with in the **estout** suite of commands. It is also possible to have variable names replaced by variable labels through the **label** option.

Table 4.1 Multilevel regression of standardized reading scores on gender, parental occupational status, and school characteristics.

	(1) z_read	(2) z_read	(3) z_read
female		0.409*** (0.0165)	0.411*** (0.0162)
pos			0.0134*** (0.000479)
pcgirls			-0.0701 (0.116)
schsize100			0.0374*** (0.00594)
_cons	-0.0366 (0.0286)	-0.240*** (0.0298)	-1.234*** (0.0876)
var(Consta~) _cons	0.268*** (0.0114)	0.269*** (0.0114)	0.180*** (0.00807)
var(Residu~) _cons	0.768*** (0.00471)	0.734*** (0.00450)	0.698*** (0.00429)
N	13646	13646	13646
AIC	36055.7	35452.5	34658.4
BIC	36078.2	35482.6	34711.1
icc	0.259	0.268	0.205
ll	-18024.8	-17722.3	-17322.2
df_m	0	1	4

Standard errors in parentheses

* p<0.05, ** p<0.01, *** p<0.001

```
esttab null model2 model3, label se aic bic ///
        scalars(icc ll df_m) transform(ln*: exp(@)^2 exp(@)^2) ///
        eqlabels("" "var(Constant)" "var(Residual)" )
```

We recommend adding a **using** option to your **esttab** command so that you can open your tables in a word-processing program to aid editing. You can save your table in many formats, including .doc, .tab, .csv, and .html. We find either .rtf or .doc to be the most flexible in terms of editing because the columns are retained in a 'column format' (rather than just as an image). Below we've added the option **using Chapter4_table1.doc, replace** to the command so that Stata will save the table in the form of a Word document in your working directory.

```
esttab null model2 model3, se aic bic scalars(icc ll df_m) ///
        transform(ln*: exp(@)^2 exp(@)^2) ///
        eqlabels("" "var(Constant)" "var(Residual)" ), ///
        using Chapter4_table1.doc, replace
```

If you do this, you will see that it comes up as a link on your results window.

```
(output written to Chapter4_table1.doc)
```

If you click on the Chapter4_table1.doc link (it will be in blue unless you changed the default screen colours in Stata), it will open in the default or your selected text viewer and may look something like this (depending on your text editor):

```
---------------------------------------------------------------------
                                (1)             (2)             (3)
                        Standardiz~)    Standardiz~)    Standardiz~)
---------------------------------------------------------------------
female                                   0.409***        0.411***
                                        (0.0165)        (0.0162)

highest parental o~t                                     0.0134***
                                                        (0.000479)

proportion of girl~l                                    -0.0701
                                                        (0.116)

School size/100 st~s     0.0374***
                                        (0.00594)

Constant                -0.0366         -0.240***       -1.234***
                        (0.0286)        (0.0298)        (0.0876)
---------------------------------------------------------------------
var(Constant)
Constant                 0.268***        0.269***        0.180***
                        (0.0114)        (0.0114)        (0.00807)
---------------------------------------------------------------------
var(Residual)
Constant                 0.768***        0.734***        0.698***
                        (0.00471)       (0.00450)       (0.00429)
---------------------------------------------------------------------
Observations             13646           13646           13646
AIC                     36055.7         35452.5         34658.4
BIC                     36078.2         35482.6         34711.1
icc                      0.259           0.268           0.205
ll                     -18024.8        -17722.3        -17322.2
df_m                         0               1               4
---------------------------------------------------------------------
```

Standard errors in parentheses
* p<0.05, ** p<0.01, *** p<0.001

With a bit of editing, we produced the version in Table 4.2.

Table 4.2 Example of presenting random intercept multilevel regression models of standardized reading scores ($N = 13,646$)

Unstandardized coefficients

	Null	Null + gender	Null + gender + controls
Intercept	−0.037	−0.240***	−1.234***
Female (ref: male)		0.409***	0.411***
Parental occupational status			0.013***
Percentage girls			−0.070
School size (/100 students)			0.037***

	Null	Null + gender	Null + gender + controls
Level 1 variance	0.268	0.269	0.180
Level 2 variance	0.768	0.734	0.698
ICC	0.259	0.268	0.205
Log likelihood	–18,024.8	–17,722.3	–17,322.2

$^*p < 0.05, ^{**}p < 0.01, ^{***}p < 0.001$

Some researchers strongly prefer retaining the standard errors in parentheses either beside or below the coefficients, but to save space, we have eliminated them. A word of caution is that you should be familiar with the convention in your discipline/audience.

WHAT DO YOU TALK ABOUT?

With so much information presented, it can seem overwhelming to choose what you should talk about. It is usually not necessary to talk about *everything* reported in your table. There are different conventions by discipline, but in general, from the above example, one could note the change in the strength of the *female* coefficient (e.g. a reduction indicates that its effect is partially mediated through the added variables) as well as the significance/direction of added variables as they pertain to hypotheses and previous research. For all model-to-model change, you could comment on the change in the ICC, and any statistical evidence that suggests that improvements have been made in the model (e.g. the log likelihood or evidence from likelihood ratio test).

Assuming we had the set of hypotheses and added the variables according to previous literature and theory (as discussed above), we could say that the null model showed that differences between schools accounted for around 27% of the variance in reading scores. This is not a trivial amount, and therefore it is a good thing we are using multilevel models to capture the Level 2 variance. Once we add the dummy variable for gender, we find that being a female is positively associated with reading scores (consistent with the voluminous literature on the topic). The ICC increases from 0.259 to 0.268, indicating that the differences between schools increase only slightly once we account for gender. Once the controls are added, we can see that the coefficient for gender doesn't change much. Of the controls, parental occupational status is positively associated with reading scores, percentage of girls in the school is non-significant, and school size is positively associated. The Level 2 variance is reduced with the addition of the block of control variables – this simply means that the variance between schools is reduced once we account for all the variables in the model. This reduction in variance across schools is also reflected in the much smaller ICC in the final model (0.205). The log likelihood increases for each model suggest that each successive model has been improved upon.

MODELS WITH RANDOM COEFFICIENTS

Until now, our examples have been focused on random intercept models. It may, however, be the case that you wish for one (or more) of your coefficients to be allowed to vary, as well as your intercepts. Remember, you should have sufficient theoretical reasoning for wanting to do this. There is an important detail that you need to be aware of about how **esttab** works with random coefficients in **mixed**, however.

Suppose you run the following commands:

```
eststo drop *

mixed z_read || schoolid:

estat icc

estadd scalar icc = r(icc2)

estimates store null

mixed z_read female || schoolid:

estat icc

estadd scalar icc = r(icc2)

estimates store model2

mixed z_read female pos escs schsize100 || schoolid:

estat icc

estadd scalar icc = r(icc2)

estimates store model3

mixed z_read female pos escs schsize100 || schoolid: ///
  escs, cov(uns) nolog

estat icc

estadd scalar icc = r(icc2)

estimates store model4
```

Here we start with the command **eststo drop**, which drops the previous model estimates from memory. We then run the null model, one with the *female* dummy variable, one with controls, then one with a random slope for the variable for family capital, *escs*.

```
esttab null model*, not transform(ln*: exp(@)^2 exp(@)^2)
```

The **esttab** command produces the table below. Note that we have asked for exponentiated standard deviations so we can make sense of the variances, as in the previous example.

Table 4.3 Example of using esttab to present nested multilevel models

	(1) z_read	(2) z_read	(3) z_read	(4) z_read
z_read				
female		0.409***	0.409***	0.407***
pos		0.00200*	0.00226**	
escs			0.316***	0.310***
schsize100			0.0322***	0.0290***
_cons	−0.0366	−0.240***	−0.680***	−0.656***
lns1_1_1				
_cons	0.268***	0.269***	0.155***	0.0173***
lnsig_e				
_cons	0.768***	0.734***	0.683***	0.676***
lns1_1_2				
_cons				0.159***
atr1_1_1_2				
_cons				−0.510***
N	13646	13646	13644	13644

*p<0.05, **p<0.01, *** p<0.001

You might look at this table and think it looks pretty good, but there is an important detail that you must note. The problem is that the ordering of the *lns1_1_1* changes meaning/order once random coefficients are added to the model. In the first three models *lns1_1_1* is the random intercept variance, but in the fourth model *lns1_1_1* is the random slope variance for the variable *escs*. Essentially *lns1_1_1* and *lns1_1_2* need to switch places in the last column in order to present the models correctly. *This is very easy to miss unless you know to look for it.* It is easiest just to switch them via a quick copy and paste, but you can use code to produce the output in Table 4.4:

```
esttab null model*, not transform(#*: exp(@)^2 exp(@)^2) ///

        equations(2:2:2:3, 3:3:3:5, .:.:.:2, .:.:.:4)
```

Table 4.4 Using options with esttab to ensure correct positioning of coefficients in nested multilevel models

	(1) z_read	(2) z_read	(3) z_read	(4) z_read
z_read				
female		0.409***	0.409***	0.407***
pos			0.00200*	0.00226**
escs			0.316***	0.310***
schsize100			0.0322***	0.0290***
_cons	−0.0366	−0.240***	−0.680***	−0.656***
#1				
_cons	0.268***	0.269***	0.155***	0.159***
#2				
_cons	0.768***	0.734***	0.683***	0.676***
#3				
_cons				0.0173***
#4				
_cons				−0.360***
N	13646	13646	13644	13644

*p<0.05, **p<0.01, *** p<0.001

The **equations** option in **esttab** allows you to reorder the different equations in your model. If you look at your output for the four models just presented, you can see that the results for the variances are presented as:

Model 1:   var(_cons), var(Residual)

Model 2:   var(_cons), var(Residual)

Model 3:   var(_cons), var(Residual)

Model 4:   var(escs), var(_cons), corr(escs, _cons), var(Residual)

The fourth model is the one where the pattern changes. In the **equations** option above:

2:2:2:3 tells Stata that the values for_cons are held in the second equation and in the second position in the results (the first position is for the fixed coefficients part of the output) for all models except the fourth, where it is pushed down one position to third. The colon between each number is necessary to tell Stata to move to the next model.

3:3:3:5 tells Stata that the values for var(Residual) are held in the third position for all models, except the last one, where it is pushed down two positions to fifth.

.:.:.:2 tells Stata that the values for var(escs) don't exist except in the final model and in the second position. The parts that look like Morse code indicate that there is nothing (.) in the model of comparable value.

.:.:.:4 tells Stata that the values for corr(escs, _cons) don't exist except in the final model and in the fourth position.

Obviously, this can be a bit confusing and it might just be easier to cut and paste, but if you construct a good do-file and want to change the fixed part of your models, this addition to your do-file might make your life a bit easier.

Of course, you'd want to fix up your output by adding more options to the **esttab** command:

```
esttab null model* using Chapter4_table4.doc, replace ///
        not scalars(icc ll) ///
        transform(#*: exp(@)^2 exp(@)^2) ///
        equations(2:2:2:3, 3:3:3:4, .:.:.:2) stardrop(#*:)
```

The option **stardrop(#*:)** removes the significance stars from your variance components. Some researchers report the significance levels of their variance components and others do not. We are of the position that it is not necessary to do so.

Table 4.5 Example of how to present random coefficient models

Unstandardized coefficients

	Null	Gender	Gender + controls	+ random coefficient for school
Intercept	−0.037	−0.240***	−0.680***	−0.656***
Female (ref: male)		0.409***	0.409***	0.407***
Parental occupational status			0.002*	0.002**
ESCS			0.316***	0.310***
School size (/100 students)			0.032***	0.029***
Variance components				
Level 1 variance	0.268	0.269	0.155	0.159
Level 2 variance	0.768	0.734	0.683	0.360
Random slope ESCS				0.017
Correlation of random slope ESCS and intercept				−0.196***
N	13646	13646	13644	13644
ICC	0.259	0.268	0.185	0.190
Log likelihood	−18024.8	−17722.3	−17158.4	−17135.3

*$p < 0.05$, **$p < 0.01$, ***$p < 0.001$

After some cosmetic edits, you could come up with something like Table 4.2. It is convention that the variance components (or different random parts of your model) are all presented in one section, and Table 4.2 gives an example of how you might go about this. You should note that there are several conventions for this type of presentation – be familiar with the convention in your discipline by looking at current journal articles.

You would talk about the fixed part of the model just as you had in the random intercept model. In terms of the random coefficient you have added to the estimation, you will note that the correlation between the random parts of the coefficient and the intercept is negative, which means that schools with higher intercepts have smaller coefficients for the independent variable. Also mention any likelihood ratio tests and whether they show an improvement in the fit of the model.

However, if you include this or any random slope in a model, you should have a theoretical reason for believing that the effect of a student's position on the economic, social and cultural status index would differentially impact upon reading scores depending on the school he or she attends. When discussing this result, you would refer back to this theoretical justification, either supporting or not supporting the hypothesis that you crafted around the predicted relationship.

WHAT ABOUT GRAPHS?

We've suggested presenting graphs where possible. This suggestion applies not only to multilevel models, but also to statistics in general. Graphs are more inviting to look at than huge tables full of numbers, and many people find graphs more accessible. Plus, they are kind of fun to make.

We have included quite a few graphs in this book already, but most of these are for the sole purpose of exploring your data and understanding multilevel models. Don't put a caterpillar plot or plots of residuals in your paper! They are neat to look at, but they are diagnostic tools. Researchers using growth curve models (a type of multilevel model) will use a very specific type of graph to display their results, but that topic is beyond the scope of this book.

We explored graphing interactions in Chapter 2, where we covered two cross-level interactions: gender (Level 1) by proportion of girls in the school (Level 2); and parental occupational status (Level 1) by teacher student ratio (Level 2). We return to this here to give you some more examples.

Interactions between your independent variables are useful to graph because their relationships can be difficult to eyeball. To review, an interaction tells you if a relationship between X_1 and Y is dependent on the level of X_2. The X values (independent variables) can be at any level. You can test characteristics of individuals on individuals (Level 1 * Level 1), or individuals on schools (Level 1 * Level 2), or schools on schools (Level 2 * Level 2). You should have good reason to do so, not just be engaged in a data mining exercise (at least, that is our strong opinion on the matter). Interaction terms change the way the coefficients for the main effects are understood, and we cover this in detail in Chapter 2.

In our first example, we interact a Level 1 variable with a Level 1 variable. We hypothesize that the effect of being female on reading scores depends on the level of parental occupational status. You may have read some literature that suggests that the gap between boys and girls closes for students with parents who have high-status jobs, for example. There are two main ways of investigating this potential interaction with one dichotomous variable (*female*) and one continuous variable (*pos*). First, include the interaction term, using factor variables, in the model. In this case we will continue with a random intercept model with gender (*female*), parental occupational status (*pos*), family capital (*escs*) and school size (*schsize100*) as independent variables:

```
mixed z_read female##c.pos escs schsize100 || schoolid:
```

The double hashtags between *female* and *c.pos* tell Stata to include the main effects for *female* and *pos* as well as the interaction terms between them. Be sure to put two hashtags (not just one), or you are given only the interaction term with no main effects (which makes no sense to us). The c. before *pos* tells Stata that *pos* is a continuous variable. If you don't add the c. it won't run because it thinks you have a categorical ('factor') variable with too many values. We obtain the following (partial) output:

```
-------------------------------------------------------------------------------
     z_read |    Coef.    Std. Err.      z     P>|z|     [95% Conf.   Interval]
------------+------------------------------------------------------------------
   1.female |   .4219721   .0490753     8.60    0.000    .3257863     .518158
        pos |   .0021181   .0009041     2.34    0.019    .000346      .0038902
            |
female#c.pos|
          1 |  -.0002488   .0008967    -0.28    0.781   -.0020064     .0015087
            |
       escs |   .3160706   .0175126    18.05    0.000    .2817466     .3503947
 schsize100 |   .0321827   .0055312     5.82    0.000    .0213418     .0430236
      _cons |  -.6861764   .0705377    -9.73    0.000   -.8244277    -.547925
-------------------------------------------------------------------------------
```

In the output, the interaction term is female#c.pos, which has a z value of –0.28 and a p value of 0.78. From this we would conclude that the interaction between gender and parental occupational status is non-significant so that the effect of parental occupational status on reading scores does not vary by gender.

Even though the above interaction is non-significant, the second way of exploring this interaction would be to convert the parental occupational status (*pos*) variable to quartiles first so we can divide up the distribution of *pos* into groups containing 25% of the values:

```
xtile quartpos=pos, nq(4)
```

The **xtile** command creates a new variable *quartpos* with values 1 to 4, where 1 is equal to the lowest quartile of *pos* scores and 4 is equal to the highest quartile of *pos* scores. Next we include an interaction term in the model:

```
mixed z_read female##quartpos escs schsize || schoolid:
```

We are presented with the following results (partial output):

```
------------------------------------------------------------------------
         z_read |    Coef.   Std. Err.    z    P>|z|   [95% Conf. Interval]
----------------+-------------------------------------------------------
       1.female |  .4287775  .0261044  16.43  0.000    .3776139    .4799411
                |
       quartpos |
              2 |  .0893324  .0300022   2.98  0.003    .0305291    .1481356
              3 |   .083363  .0318449   2.62  0.009    .0209482    .1457778
              4 |   .102409  .0359103   2.85  0.004    .0320261    .1727919
                |
female#quartpos |
            1 2 | -.0088223  .0402619  -0.22  0.827   -.0877341    .0700895
            1 3 | -.0833696  .0395208  -2.11  0.035   -.1608288   -.0059103
            1 4 |  .0048637   .039818   0.12  0.903   -.0731781    .0829056
                |
           escs |  .3166568   .01546   20.48  0.000    .2863558    .3469578
     schsize100 |  .0321323  .0055284   5.81  0.000    .0212968    .0429678
          _cons | -.6367444  .0565048 -11.27  0.000   -.7474916   -.5259971
------------------------------------------------------------------------
```

As we are testing for interactions, at this point all we want to do is look to see if the interactions are significant. If they aren't, there is no point in going to additional lengths to graph them! You can see that the main effects for *female* and the different values of *quartpos* are statistically significant from their respective reference categories: male and the lowest quartile of parental occupational status. The results for the interaction term *female*quartpos* show the coefficients between *female* and categories 2, 3, and 4 of *quartpos*. Thus, *female* = 0 and *quartpos* = 1 are the reference categories. Stata automatically chooses the lowest coded category of a variable to make the reference category.

If we had wanted the fourth category of *quartpos* to be the reference category, we could have changed the reference category by typing **female##b4. quartpos**. The interaction between *female* and the third quartile of *quartpos* is significant and negative. It is rather hard to determine exactly what this means if we are tasked with adding up all the main effects and consider the reference categories. Fortunately, Stata has the command **margins** to do this for us.

```
margins female#quartpos
```

In this step, the marginal effects of the interaction are calculated, holding the rest of the values in the model at their mean values. Sometimes this might make sense and you can change the values of other variables in the options in the **margins** command. However, for the purposes of this example, it is fine. The margins command brings up a table:

```
Predictive margins                      Number of obs = 13644
Expression : Linear prediction, fixed portion, predict()

------------------------------------------------------------------------
                 |            Delta-method
                 |   Margin   Std. Err.     z    P>|z|   [95% Conf. Interval]
-----------------+------------------------------------------------------
female#quartpos  |
            0 1  |  -.2785425  .0300605   -9.27  0.000   -.3374599   -.2196251
            0 2  |  -.1892102  .030633    -6.18  0.000   -.2492497   -.1291706
            0 3  |  -.1951796  .0302468   -6.45  0.000   -.2544622   -.135897
            0 4  |  -.1761335  .0320201   -5.50  0.000   -.2388917   -.1133754
            1 1  |   .150235   .0301848    4.98  0.000    .0910738    .2093961
            1 2  |   .230745   .0305525    7.55  0.000    .1708632    .2906269
            1 3  |   .1502284  .030344     4.95  0.000    .0907553    .2097015
            1 4  |   .2575077  .0324979    7.92  0.000    .193813     .3212024
------------------------------------------------------------------------
```

This table itself is not tremendously useful alone. But it is a necessary first step to getting the graph through the next command, **marginsplot**.

```
marginsplot, recast(line) recastci(rarea)xdimension(quartpos)
```

If you just type **marginsplot** with no options, you get a graph, but not one that is particularly useful in this situation. The default is a line chart with *female* on the *x*-axis. We add some options to get an area chart (graphing the confidence intervals around the estimate) and place *quartpos* on the *x*-axis, having two separate lines for each value of *female* (see Figure 4.1).

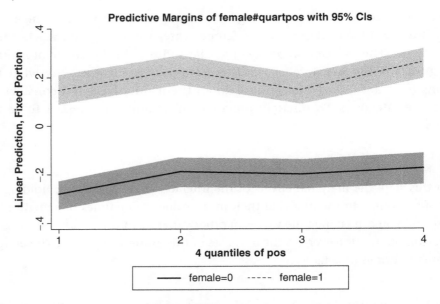

Figure 4.1 Predictive margins plot of gender and parental occupational status interaction

Of course you'd want to open the graph editor and edit it properly (titles, axes, legend), but this is what you get without any editing. You can see that the interaction between *quartpos* at the third quartile and *female* demonstrates a narrowing of the reading scores between males and females at the third quartile, but this spreads out again at the highest quartile of *pos*. Thus it appears to be within the middle to upper-middle class that *pos* has the effect of narrowing the gap between males and females.

In terms of where to add this interaction in a table, the **esttab** command will add this line to your output table, just as in other models. It just treats it as more variables in the 'fixed' part of your output.

CROSS-LEVEL INTERACTIONS

Multilevel modeling also allows researchers to examine if individual characteristics impact upon an outcome of interest depending on a Level 2 variable. For example, we may have reason to believe that the impact of being a female on reading scores is dependent upon school size. Perhaps in larger schools, the positive effect of being female on reading scores diminishes. These cross-level interactions are no different to model than the Level 1 by Level 1 interactions (or Level 2 by Level 2 interactions).

We add the interaction of *female* and *schsize100* to the model. The hashtags between the two variables indicate that it is an interaction to include main effects, and the **c.** in front of *schsize100* tells Stata that it is a continuous variable.

```
mixed z_read female##c.schsize100 escs pos|| schoolid:
```

```
------------------------------------------------------------------------
        z_read |   Coef.    Std. Err.    z    P>|z|   [95% Conf. Interval]
---------------+--------------------------------------------------------
      1.female |  .4678149  .0387786   12.06  0.000   .3918103    .5438196
     schsize100 |  .0353757  .0058511    6.05  0.000   .0239077    .0468437
               |
female#c.schsize100 |
             1 | -.0065842  .0039692   -1.66  0.097  -.0143637    .0011952
               |
          escs |  .3164181  .01751211   8.07  0.000   .2820951    .3507411
           pos |  .0019816  .0007919    2.50  0.012   .0004295    .0035336
         _cons | -.7082947  .0690151  -10.26  0.000  -.8435617   -.5730276
------------------------------------------------------------------------
```

The coefficient for the interaction term *female#c.schsize100* is only statistically significant at the 0.10 level here, which is kind of a stretch, given that our sample size is over 13,000. It would probably be a good idea just to say that your hypothesis isn't supported, or is only very weakly supported. However, to follow through with how to graph this result, you have to ask Stata for the margins.

```
margins, at(schsize100=(1(5)26)) over(female)
```

Because *schsize100* is a continuous variable, you must use the **at** option in order for Stata to render this graph. Here, we tell Stata to graph *schsize100* at 5-unit increments (i.e. 500 students), starting at 1 and ending at 26.

After margins have been calculated, you may request

```
marginsplot
```

With no options specified, you get Figure 4.2. We can see that there is only a small narrowing of the gap between males and females as the school size increases.

In terms of adding cross-level interactions to a table of results, **esttab** will present the results in the fixed portion of your estimation. In this example we use the **wide** and **se** options, which tell Stata to put the standard error to the right (wide format) of the coefficient. If **wide** is not used then the standard error is reported blow the coefficient.

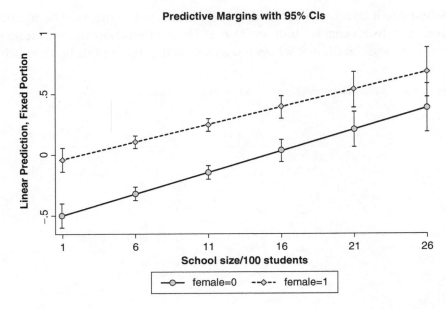

Figure 4.2

```
mixed z_read female##c.schsize100 escs pos || schoolid:

estimates store xlevelinteraction

estat icc

estadd scalar icc = r(icc2)

esttab xlevelinteraction, wide se scalars(icc ll), ///
        using Chapter4_table3.doc, replace
```

The Chapter4_table6.doc link produces something like the following:

	(1) z_read	
z_read		
0.female	0	(.)
1.female	0.468	(0.0388)
schsize100	0.0354	(0.00585)
0.female~100	0	(.)
1.female~100	−0.00658	(0.00397)
escs	0.316	(0.0175)
pos	0.00198	(0.000792)
_cons	−0.708	(0.0690)
lns1_1_1		
_cons	0.155	(0.00709)
lnsig_e		
_cons	0.683	(0.00420)

```
N                           13644
icc                         0.185
ll                        -17157.0
```

```
Standard errors in parentheses
* p<0.05, ** p<0.01, *** p<0.001
```

With some cosmetic editing you can get a table like Table 4.6.

Table 4.6 Example of linear multilevel model of standardized reading scores regressed on sex, school size, controls and cross-level interaction ($N = 13,644$)

	B	SE
Intercept	−0.708***	(0.069)
Female (ref: male)	0.468***	(0.039)
School size (/100 students)	0.035***	(0.000)
Family capital	0.316***	(0.018)
Parent occupational status	0.002*	(0.001)
Cross-level interaction		
Female * school size	−0.007*	(0.004)
Variance Components		
Level 1 variance	0.155	(0.007)
Level 2 variance	0.683	(0.004)
ICC	0.185	
Log likelihood	−17157.0	

Standard errors in parentheses
$^{+}p < 0.10, ^{*}p < 0.05, ^{**}p < 0.01, ^{***}p < 0.001$

You would discuss it in similar terms as you did above, additionally addressing the hypotheses around the cross-level interaction. A very good and simply explained example of this type of discussion of the findings of cross-level interactions is found in Huijts and Kraaykamp (2012).

PARTING WORDS

In this brief book, we have tried to acquaint you with the world of multilevel models by giving you a taster of what these models are used for and how to use them.

In this final chapter, we have shown you how to present your results in a way that is most informative to your research audience. As with all techniques that lend themselves to easy expansion and complication through the addition of a few options here and there, we remind the reader (for the final time, promise!) that it can be easy to lose sight of your original research question once you've seen all that *could* be done. More complicated isn't always better. The best way to become comfortable with these techniques is to see what others *in your field* are doing and to take a lead from their techniques. We emphasize 'in your field' because there are discipline-specific practices that are not necessarily appropriate to other fields. Read widely and you will see that economists, sociologists, epidemiologists and education researchers have widely different practices for reporting on their methods, even if they've done essentially the same thing. There are many reasons for this (historical development of the discipline and the reliance on discipline-specific theories, for example), but it doesn't change the fact that everyone is just trying to answer their research question with the data that they have. This is just another technique to add to your methodological toolkit. And the more you have, the more likely you are to answer research questions in the most appropriate way.

CHAPTER 4 TAKEAWAY POINTS

- It is important to present results in a manner that is most appropriate for your target audience, which can vary considerably by discipline.
- You should give some mention to the results of your null model to demonstrate that multilevel modeling was necessary.
- In this chapter, we have recommended five steps to be followed in model-building.
- It is OK to focus your discussion on the fixed part of your model if your research questions are not specifically concerned with the random parts; that is, you are using random intercepts or coefficients solely to properly specify your model.
- Wherever possible, you should include graphic representations of your findings to make them easier to understand.

References

Albright, J.J. and Marinova, D.M. (2010) Estimating multilevel models using SPSS, Stata, SAS, and R. Technical Working Paper, University Information and Technology Service (UITS) Center for Statistical and Mathematical Computing, Indiana University. Open Access. Available at: http://www.indiana.edu/~statmath/stat/all/hlm/hlm.pdf

Australian Government (no date) Australian education system? Available at: http://www.studyinaustralia.gov.au/global/australian-education/education-system (accessed 31May 2015).

Bell, J. (2002) Analysing student progress in higher education using cross-classified multilevel logistic models. Society for Multivariate Analysis in the Behavioural Sciences, Tilburg, The Netherlands, July. Open Access. Available at: http://www.cambridgeassessment.org.uk/Images/109705-analysing-student-progress-in-higher-education-using-cross-classified-multilevel-logistic-models.pdf

Brambor, T., Clark, W.R. and Golder, M. (2006) Understanding interaction models: Improving empirical analyses. *Political Analysis*, 14(1): 63–82. Open Access. Available at: http://pan.oxfordjournals.org/content/14/1/63.full.pdf+html

Bronfenbrenner, U. (1977) Toward an experimental ecology of human development. *American Psychologist*, 32(7): 513–31.

Bronfenbrenner, U. (2001) The bioecological theory of human development. In N.J. Smelser and P.B. Baltes (eds), *International Encyclopedia of the Social and Behavioural Sciences* (Vol. 10, pp. 6963–70). New York: Elsevier.

Bryan, M. and Jenkins, S.J. (2013) Regression analysis of country effects using multilevel data: A cautionary tale. ISER Working Paper Series 2013-14. Open Access. Available at: https://www.iser.essex.ac.uk/research/publications/working-papers/iser/2013-14

Bryk, A.S. and Raudenbush, S.W. (2002) *Hierarchical Linear Models in Social and Behavioral Research: Applications and Data Analysis Methods* (2nd edn). Thousand Oaks, CA: Sage.

Clarke, P., Crawford, C., Steele, F. and Vignoles, A. (2013) Revisiting fixed- and random-effects models: Some considerations for policy-relevant education research. *Education Economics*, 23(3), 259–77. doi: 10.1080/09645292.2013.855705

Courgeau, D. (2003) *Methodology and Epistemology of Multilevel Analysis: Approaches from Different Social Sciences*. Dordrecht: Kluwer Academic.

Diez-Roux, A.V. (2000) Multilevel analysis in public health research. *Annual Review of Public Health*, 21: 171–92. doi: 10.1146/annurev.publhealth.21.1.171

Duncan, C., Jones, K. and Moon, G. (1998) Context, composition, and heterogeneity: Using multilevel models in health research. *Social Science and Medicine*, 46(1): 97–117. doi: 10.1016/S0277-9536(97)00148-2

Enders, C.K. and Tofighi, D. (2007) Centering predictor variables in cross-sectional multilevel models: A new look at an old issue. *Psychological Methods*, 12(2): 121–38. doi: 10.1037/1082-989X.12.2.121

Fitzmaurice, G.M., Laird, N.M. and Ware, J. H. (2012) *Applied Longitudinal Analysis*. Hoboken, NJ: John Wiley & Sons.

Gelman, A. (2006) Prior distributions for variance parameters in hierarchical models. *Bayesian Analysis*, 1(3): 515–33. doi: 10.1214/06-BA117A Open Access. Available at: http://projecteuclid.org/download/pdf_1/euclid.ba/1340371048

Gorard, S. (2003a) What is multi-level modelling for? *British Journal of Educational Studies*, 51(1): 46–63. doi: 10.1111/1467-8527.t01-2-00224

Gorard, S. (2003b) In defence of a middle way: A reply to Plewis and Fielding. *British Journal of Educational Studies*, 51(4): 420–6. doi: 10.1046/j.1467-8527.2003.00247.x

Gorard, S. (2007) The dubious benefits of multi-level modeling. *International Journal of Research & Method in Education*, 30(2): 221–36. doi: 10.1080/17437270701383560

Halaby, C. N. (2004) Panel models in sociological research: theory into practice. *Annual Review of Sociology*, 30: 507–44. doi: 10.1146/annurev.soc.30.012703.110629

Hox, J.J. (1998) Multilevel modeling: When and why. In I. Balderjahn, R. Mathar, and M. Schader (eds), *Classification, Data Analysis and Data Highways* (pp. 147–54). Berlin: Springer.

Hox, J.J. (2010). *Multilevel Analysis. Techniques and Applications* (2nd edn.). New York, NY: Routledge.

Hox, J.J. and Kreft, I.G.G. (1994) Multilevel analysis methods. *Sociological Methods and Research*, 22(3): 283–99. doi: 10.1177/0049124194022003001

Huijts, T. and Kraaykamp, G. (2012) Formal and informal social capital and self-rated health in Europe: A new test of accumulation and compensation mechanisms using a multi-level perspective. *Acta Sociologica*, 55(2), 143–58. doi: 10.1177/0001699312439080

Kreft, I.G.G. and de Leeuw, J. (1998) *Introducing Multilevel Modeling*. London: Sage.

Langford, I.H. and Lewis, T. (1998) Outliers in multilevel data. *Journal of the Royal Statistical Society, Series A*, 161: 121–60. doi: 10.1111/1467-985X.00094

Leckie, G. (2012) Cross-classified multilevel models using Stata: How important are schools and neighbourhoods for students' educational attainment?

In D. Garson (ed.), *Hierarchical Linear Modeling: Guide and Applications* (pp. 311–32). Thousand Oaks, CA: Sage.

Leckie, G. (2013) Cross-classified multilevel models – Stata practical. LEMMA VLE module 12, 1–52. http://www.bristol.ac.uk/cmm/learning/course.html

Lee, V.E. (2000) Using hierarchical linear modeling to study social contexts: The case of school effects. *Educational Psychologist*, 35(2): 125–41. doi: 10.1207/S15326985EP3502_6

Maas, C.J.M. and Hox, J.J. (2005) Sufficient sample sizes for multilevel modeling. *Methodology*, 1(3): 86–92. doi: 10.1027/1614-2241.1.3.86

Mok, M. (1995) Sample size requirements for 2-level designs in educational research. *Multilevel Modelling Newsletter*, 7(2), 11–15.

Nezlek, J.B. (2008) An introduction to multilevel modeling for social and personality psychology. *Social and Personality Psychology Compass*, 2(2): 842–60. doi: 10.1111/j.1751-9004.2007.00059.x

OECD (2009) *ISA 2006 Technical Report*. Paris: OECD. Open Access. Available at: http://www.oecd.org/pisa/pisaproducts/42025182.pdf

Pevalin, D. and Robson, K. (2009) *The Stata Survival Manual*. Maidenhead: Open University Press.

Pike, G.R. and Rocconi, L.M. (2012) Multilevel modeling: Presenting and publishing the results for internal and external constituents. In J.L. Lott and J.S. Antony (eds), *Multilevel Modeling Techniques and Applications in Institutional Research*, New Directions for Institutional Research No. 154 (pp. 111–24). San Francisco: Jossey-Bass.

Plewis, I. and Fielding, A. (2003) What is multi-level modelling for? A critical response to Gorard. *British Journal of Educational Studies*, 51(4): 408–19. doi: 10.1046/j.1467-8527.2003.00246.x

Rabe-Hesketh, S. and Skrondal, A. (2008) *Multilevel and Longitudinal Modeling Using Stata* (2nd edn). College Station, TX: Stata Press.

Rabe-Hesketh, S. and Skrondal, A. (2012) *Multilevel and Longitudinal Modeling Using Stata* (3rd edn). College Station, TX: Stata Press.

Ranstam, J. (2009) Sampling uncertainty in medical research. *Osteoarthritis and Cartilage*, 17(11): 1416–19. Open Access. Available at: http://www.sciencedirect.com/science/article/pii/S1063458409001071

Richter, T (2006). What is wrong with ANOVA and multiple regression? Analyzing sentence reading times with hierarchical linear models. *Discourse Processes*, 41: 221–50. doi: 10.1207/s15326950dp4103_1

Snijders, T.A.B. (2005) Power and sample size in multilevel linear models. In B.S. Everitt and D.C. Howell (eds), *Encyclopedia of Statistics in Behavioral Science* (Vol. 3, pp. 1570–73). Hoboken, NJ: John Wiley & Sons.

Snijders, T.A.B. and Bosker, R.J. (1994) Modeled variance in two-level models. *Sociological Methods & Research*, 22(3): 342–63. doi: 10.1177/0049124194022003004

Snijders, T.A.B. and Bosker, R.J. (1999) *Multilevel Analysis: An Introduction to Basic and Applied Multilevel Analysis.* London: Sage.

Tate, R.L. (2004) A cautionary note on shrinkage estimates of school and teacher effects. *Florida Journal of Educational Research*, 42: 1–21. Retrieved from http://www.coedu.usf.edu/fjer/2004/FJERV42P0121.pdf

Index

Tables and Figures are indicated by page numbers in bold print. Page numbers in italics indicate Stata commands